Lecture Notes in Computer Science 12417

More information about this series at http://www.springer.com/series/7412

Ninon Burgos · David Svoboda ·
Jelmer M. Wolterink · Can Zhao (Eds.)

Simulation and Synthesis in Medical Imaging

5th International Workshop, SASHIMI 2020
Held in Conjunction with MICCAI 2020
Lima, Peru, October 4, 2020
Proceedings

Editors
Ninon Burgos
CNRS – Paris Brain Institute
Paris, France

David Svoboda
Masaryk University
Brno, Czech Republic

Jelmer M. Wolterink
University of Twente
Enschede, The Netherlands

Can Zhao
Johns Hopkins University
Bethesda, MD, USA

ISSN 0302-9743 ISSN 1611-3349 (electronic)
Lecture Notes in Computer Science
ISBN 978-3-030-59519-7 ISBN 978-3-030-59520-3 (eBook)
https://doi.org/10.1007/978-3-030-59520-3

LNCS Sublibrary: SL6 – Image Processing, Computer Vision, Pattern Recognition, and Graphics

This Springer imprint is published by the registered company Springer Nature Switzerland AG
The registered company address is: Gewerbestrasse 11, 6330 Cham, Switzerland

Preface

The Medical Image Computing and Computer Assisted Intervention (MICCAI) community needs data with known ground truth to develop, evaluate, and validate computerized image analytic tools, as well as to facilitate clinical training. Since synthetic data are ideally suited for this purpose, a full range of models underpinning image simulation and synthesis, also referred to as image translation, cross-modality synthesis, image completion, etc., have been developed over the years: (a) machine and deep learning methods based on generative models; (b) simplified mathematical models to test segmentation, tracking, restoration, and registration algorithms; (c) detailed mechanistic models (top–down), which incorporate priors on the geometry and physics of image acquisition and formation processes; and (d) complex spatio-temporal computational models of anatomical variability, organ physiology, and morphological changes in tissues or disease progression.

The goal of the Simulation and Synthesis in Medical Imaging (SASHIMI)[1] workshop is to bring together all those interested in such problems in order to engage in invigorating research, discuss current approaches, and stimulate new ideas and scientific directions in this field. The objectives are to (a) bring together experts on image synthesis to raise the state of the art; (b) hear from invited speakers outside of the MICCAI community, for example in the areas of transfer learning, generative adversarial networks, or variational autoencoders, to cross-fertilize these fields; and (c) identify challenges and opportunities for further research. We also want to identify the suitable approaches to evaluate the plausibility of synthetic data and to collect benchmark data that could help with the development of future algorithms.

The 5th SASHIMI workshop was successfully held in conjunction with the 23rd International Conference on Medical Image Computing and Computer-Assisted Intervention (MICCAI 2020) as a satellite event on October 4, 2020. Submissions were solicited via a call for papers circulated by the MICCAI organizers, via the social media, as well as by directly emailing colleagues and experts in the area. Each of the 27 submissions received underwent a double-blind review by at least two members of the Program Committee, consisting of researchers actively contributing in the area. Compared to the 2019 edition, we saw an increased number of submissions and diversity of covered topics. At the conclusion of the review process, 19 papers were accepted. Overall, the contributions span the following broad categories in alignment with the initial call for papers: methods based on generative models or adversarial learning for MRI/CT/microscopy image synthesis, and several applications of image synthesis and simulation for data augmentation, image enhancement, or segmentation.

We would like to thank everyone who contributed to this 5th workshop: members of the Organizing Committee for their assistance, the authors for their contributions, the members of the Program Committee for their precious review works, promotions of the

[1] http://www.sashimi.aramislab.fr.

workshop, and general support. We are particularly grateful to the invited speaker, Dr. Ming-Yu Liu, who kindly shared his expertise and knowledge with the community. We also thank the members of the Steering Committee for their advice and support, and the MICCAI society for the general support.

October 2020

Ninon Burgos
David Svoboda
Jelmer M. Wolterink
Can Zhao

Organization

Program Chairs

Steering Committee

Program Committee

Contents

Contrast Adaptive Tissue Classification by Alternating Segmentation and Synthesis 1
Dzung L. Pham, Yi-Yu Chou, Blake E. Dewey, Daniel S. Reich, John A. Butman, and Snehashis Roy

3D Brain MRI GAN-Based Synthesis Conditioned on Partial Volume Maps 11
Filip Rusak, Rodrigo Santa Cruz, Pierrick Bourgeat, Clinton Fookes, Jurgen Fripp, Andrew Bradley, and Olivier Salvado

Synthesizing Realistic Brain MR Images with Noise Control 21
Lianrui Zuo, Blake E. Dewey, Aaron Carass, Yufan He, Muhan Shao, Jacob C. Reinhold, and Jerry L. Prince

Simulated Diffusion Weighted Images Based on Model-Predicted Tumor Growth 32
Pamela R. Jackson, Andrea Hawkins-Daarud, and Kristin R. Swanson

Blind MRI Brain Lesion Inpainting Using Deep Learning 41
José V. Manjón, José E. Romero, Roberto Vivo-Hernando, Gregorio Rubio, Fernando Aparici, Maria de la Iglesia-Vaya, Thomas Tourdias, and Pierrick Coupé

High-Quality Interpolation of Breast DCE-MRI Using Learned Transformations 50
Hongyu Wang, Jun Feng, Xiaoying Pan, Di Yang, and Baoying Chen

A Method for Tumor Treating Fields Fast Estimation 60
Reuben R. Shamir and Zeev Bomzon

Heterogeneous Virtual Population of Simulated CMR Images for Improving the Generalization of Cardiac Segmentation Algorithms. 68
Yasmina Al Khalil, Sina Amirrajab, Cristian Lorenz, Jürgen Weese, and Marcel Breeuwer

DyeFreeNet: Deep Virtual Contrast CT Synthesis 80
Jingya Liu, Yingli Tian, A. Muhteşem Ağıldere, K. Murat Haberal, Mehmet Coşkun, Cihan Duzgol, and Oguz Akin

A Gaussian Process Model Based Generative Framework for Data Augmentation of Multi-modal 3D Image Volumes 90
Nicolas H. Nbonsou Tegang, Jean-Rassaire Fouefack, Bhushan Borotikar, Valérie Burdin, Tania S. Douglas, and Tinashe E. M. Mutsvangwa

Frequency-Selective Learning for CT to MR Synthesis 101
Zi Lin, Manli Zhong, Xiangzhu Zeng, and Chuyang Ye

Uncertainty-Aware Multi-resolution Whole-Body MR to CT Synthesis 110
Kerstin Kläser, Pedro Borges, Richard Shaw, Marta Ranzini, Marc Modat, David Atkinson, Kris Thielemans, Brian Hutton, Vicky Goh, Gary Cook, M. Jorge Cardoso, and Sébastien Ourselin

UltraGAN: Ultrasound Enhancement Through Adversarial Generation 120
Maria Escobar, Angela Castillo, Andrés Romero, and Pablo Arbeláez

Improving Endoscopic Decision Support Systems by Translating Between Imaging Modalities 131
Georg Wimmer, Michael Gadermayr, Andreas Vécsei, and Andreas Uhl

An Unsupervised Adversarial Learning Approach to Fundus Fluorescein Angiography Image Synthesis for Leakage Detection. 142
Wanyue Li, Yi He, Jing Wang, Wen Kong, Yiwei Chen, and GuoHua Shi

Towards Automatic Embryo Staging in 3D+t Microscopy Images Using Convolutional Neural Networks and PointNets 153
Manuel Traub and Johannes Stegmaier

Train Small, Generate Big: Synthesis of Colorectal Cancer Histology Images 164
Srijay Deshpande, Fayyaz Minhas, and Nasir Rajpoot

Image Synthesis as a Pretext for Unsupervised Histopathological Diagnosis 174
Dejan Štepec and Danijel Skočaj

Auditory Nerve Fiber Health Estimation Using Patient Specific Cochlear Implant Stimulation Models 184
Ziteng Liu, Ahmet Cakir, and Jack H. Noble

Author Index 195

Contrast Adaptive Tissue Classification by Alternating Segmentation and Synthesis

Dzung L. Pham[1,2,3](✉), Yi-Yu Chou[1,3], Blake E. Dewey[2], Daniel S. Reich[3], John A. Butman[3], and Snehashis Roy[3]

[1] Henry M. Jackson Foundation, Bethesda, MD 20892, USA
{dzung.pham,yiyu.chou}@nih.gov

[2] Johns Hopkins University, Baltimore, MD 21218, USA
blake.dewey@jhu.edu

[3] National Institutes of Health, Bethesda, MD 20892, USA
reichds@ninds.nih.gov, jbutmana@cc.nih.gov, snehashis.roy@nih.gov

Abstract. Deep learning approaches to the segmentation of magnetic resonance images have shown significant promise in automating the quantitative analysis of brain images. However, a continuing challenge has been its sensitivity to the variability of acquisition protocols. Attempting to segment images that have different contrast properties from those within the training data generally leads to significantly reduced performance. Furthermore, heterogeneous data sets cannot be easily evaluated because the quantitative variation due to acquisition differences often dwarfs the variation due to the biological differences that one seeks to measure. In this work, we describe an approach using alternating segmentation and synthesis steps that adapts the contrast properties of the training data to the input image. This allows input images that do not resemble the training data to be more consistently segmented. A notable advantage of this approach is that only a single example of the acquisition protocol is required to adapt to its contrast properties. We demonstrate the efficacy of our approaching using brain images from a set of human subjects scanned with two different T1-weighted volumetric protocols.

Keywords: Segmentation · Synthesis · Magnetic resonance imaging · Harmonization · Domain adaptation

1 Introduction

Automated segmentation algorithms for quantifying brain structure in magnetic resonance (MR) images are widely used in neuroscientific research, and increasingly being applied in clinical trials and diagnostic applications. The best performing algorithms currently rely on training data or atlases to serve as exemplars of how MR images should be segmented. However, MR image contrast is notoriously sensitive to both hardware differences (e.g., scanner manufacturers, receiver coils) and software differences (e.g., pulse sequence parameters,

N. Burgos et al. (Eds.): SASHIMI 2020, LNCS 12417, pp. 1–10, 2020.
https://doi.org/10.1007/978-3-030-59520-3_1

software platform versions). Segmentation approaches requiring training data perform suboptimally when faced with imaging data that possess contrast properties that differ from the atlases. Perhaps more importantly, these algorithms provide inconsistent results when applied to heterogeneously acquired imaging data [12,22]. Techniques for generating more consistent segmentations across different acquisition protocols are therefore needed to enable more accurate monitoring of anatomical changes, as well as for increasing statistical power [11].

Multiple approaches have been previously proposed to perform harmonization of heterogeneous imaging data. A common approach is to incorporate site effects in the statistical modeling [3,10]. Such approaches are designed for group analyses and are complementary to image-based harmonization methods. Intensity normalization techniques that attempt to align the histograms of images using linear or piecewise linear transformations have also been proposed [17,21]. Because these transformations affect the global histogram, local contrast differences and differences in the overall anatomy across images are not well addressed. Image synthesis techniques, where a set of images from a subject is used in combination with training data to create a new image with desirable intensity properties, have also been used for harmonization [14]. A recent approach used subjects scanned with multiple acquisition protocols to serve as training data for the synthesis [6]. Such training data is often not available, however.

An alternative to harmonizing the appearances of images is to use a segmentation algorithm that is robust to variations in pulse sequences. Generative and multi-atlas segmentation algorithms have been proposed [8,20], as well as deep learning algorithms employing domain adaptation [2,7,16]. The latter approaches typically employ adversarial learning that are capable of addressing contrast variations without the need for paired training data. A disadvantage of domain adaptation approaches is that they require multiple examples of the different acquisition protocols to adequately learn features invariant to the different domains. A deep learning segmentation approach that does not require groups of training data is described in [15] that employs a parametric model of MRI image formation to augment the appearance of the training data.

In this work, the appearance of the atlas MR images is altered to resemble the contrast of the input image without changing the atlas labels. This is accomplished by using alternating steps of segmentation and synthesis. A preliminary demonstration of this framework using Gaussian classifiers and a synthesis based on a linear combination of tissue memberships was previously described in [19]. The approach here employs deep learning networks for both the segmentation and synthesis processes, and more extensive evaluation results are shown to demonstrates its efficacy.

2 Methods

We denote a segmentation SEG(·) to be a mapping from an input MR image F to its anatomical labels L, which represent the desired regions of interest. We consider supervised segmentations that require a training data set consisting of MR image and label image pairs, F_A and L_A. We denote the synthesis SYN(·) to be a mapping from the labels to an MR image. Our framework, referred to as CAMELION (Contrast Adaptive Method for Label Identification), uses these dual operations to update both the desired segmentation L and the appearance of the atlas images F_A, while keeping the input image F and the atlas labels L_A fixed. Figure 1 illustrates this process. In the first iteration, the segmentation is performed as usual with the available training data, yielding an initial segmentation estimate $L^{(0)}$. Because of differences between the atlas MR images $F_A^{(0)}$ and the input image F, this segmentation will be suboptimal. To compensate, the initial segmentation and input MR image are used to train a synthesis network, which is then applied to the atlas label images to generate new atlas MR images $F_A^{(1)}$. The segmentation network is retrained with the updated atlas images, and a new segmentation of the input MRI image, $L^{(1)}$, is computed. This process is repeated until convergence.

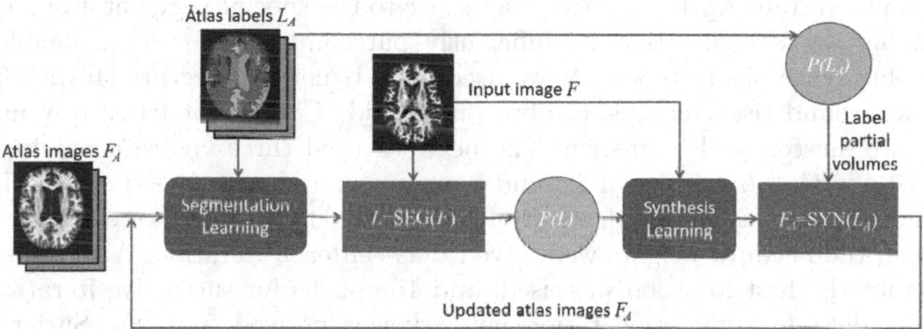

Fig. 1. Block diagram depicting the CAMELION framework. An intermediate segmentation of the input data is used to train a synthesis network, mapping atlas labels to a contrast similar to the input. Blue blocks are training, red are testing, and green are partial volume estimation. (Color figure online)

An advantage of the CAMELION framework is that only a single example of the input image's acquisition protocol is required to train the synthesis. If multiple input images with the same protocol are available, the processing could be performed group-wise, potentially with better results because of increased training data. In this work, however, we focus on processing each input image data set independently without pre-training any of the networks.

2.1 Data

Data for this study was collected under an IRB-approved protocol from 18 healthy subjects with familial relation to a person with multiple sclerosis. Subjects underwent MRI scanning with two different T1-weighted volumetric protocols on a Siemens Skyra 3T Scanner. The first scan was a Magnetization Prepared Rapid Gradient Echo (MPRAGE) protocol acquired at 1mm isotropic spatial resolution (TR = 3000 ms, TE = 3.03 ms, TI = 900 ms, FA = 9°). The second scan was a Spoiled Gradient Recalled (SPGR) protocol also acquired at 1mm isotropic resolution (TR = 7.8 ms, TE = 3 ms, FA = 18°). Although both scans are T1-weighted, the contrast properties are quite different. Both scans were rigidly co-registered [1] and then processed to remove intensity inhomogeneities [23] and extracerebral tissue [13]. Figure 2 shows examples of pre-processed data from these two acquisitions.

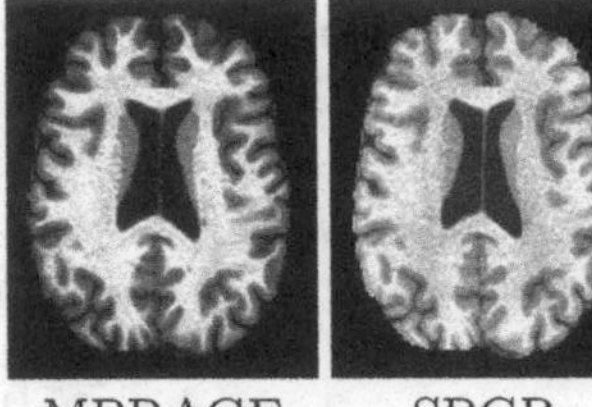

Fig. 2. T1-w images from the same subject

2.2 Segmentation

The proposed framework is relatively agnostic to the specific segmentation algorithm, but does require that it sufficiently parcellates the brain to enable a reasonable synthesis. In this work, we used a 3D U-net architecture [4] with five non-background tissue classes: cerebrospinal fluid (CSF), ventricles, gray matter, white matter, and brainstem. The network used three layers, a patch size of $64 \times 64 \times 64$, a batch size of 32, and a mean squared error loss function. The Adam optimizer was used with a learning rate of 0.001. In order to reduce training time, the network weights were saved between each iteration. A total of 30 epochs for the first iteration was used, and 15 epochs for successive iterations. Training data for the segmentation network was derived from FreeSurfer [9] applied to MPRAGE images from ten subjects, with FreeSurfer labels merged appropriately into the five aforementioned tissue classes (see Fig. 3(b)). White matter is shown in white, gray matter in gray, and CSF in dark gray. Brainstem is not present in this particular slice. Note that the CSF class in FreeSurfer does not include sulcal CSF, so was excluded from the evaluation results.

2.3 Partial Volume Estimation

Although it is possible to synthesize an image directly from discrete tissue labels, we have empirically found that synthetic images were more realistic using continuous labels, such as spatial probability functions or partial volume estimates. The advantage of continuous labels is further increased because the synthesis network is trained with a single MR image and segmentation pair. We therefore apply a partial volume estimation step following segmentation to generate a

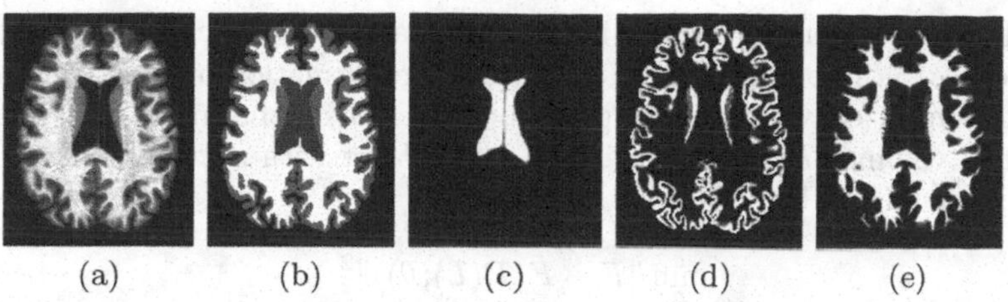

Fig. 3. Axial slices from an atlas data set: (a) MPRAGE image, (b) MPRAGE segmentation using FreeSurfer, (c) ventricle partial volume, (d) gray matter partial volume, (e) white matter partial volume. Partial volume images were computed from the MPRAGE and FreeSurfer segmentation and are used for image synthesis.

continuous function from zero to one for tissue class. For each voxel j, the input image intensity f_j is modeled as

$$f_j = \sum_{k=1}^{K} p_{jk} c_k + \eta_j \tag{1}$$

where p_{jk} is the partial volume of class k at j, c_k is the mean intensity of tissue class k, and η_j is a Gaussian noise term. In addition to the constraints that $p_{jk} > 0$ and $\sum_{k=1}^{K} p_{jk} = 1$, we restrict partial volumes to only be greater than zero for a maximum of two classes. The two non-zero classes are determined by the associated discrete segmentation of the image as the tissue classes at the voxel and the spatially nearest tissue class different from the original class. The mean intensity for each is directly estimated from the discrete segmentation and input image.

We further impose a prior probability distribution on p_{jk} such that:

$$p_{jk} = \begin{cases} \frac{1}{Z} \exp\left(\beta (p_{jk} - 0.5)^2\right) & p_{jk} \in [0, 1] \\ 0 & \text{otherwise} \end{cases} \tag{2}$$

where Z is a normalizing constant, and β is a weighting parameter. This prior allows the model to favor pure tissue classes over partial volume tissues for positive values of β. A maximum a posteriori estimate of p_{jk} can be straightforwardly computed for every j using this model under the provided constraints. In this work, β was determined empirically based on the visual quality of the partial volumes and then held fixed for all experiments. Figures 3(c)–(e) show an example result computed from the MPRAGE image in Fig. 3(a) and its segmentation in Fig. 3(b).

2.4 Synthesis

The synthesis process in this framework maps a set of tissue partial volumes back to the MR image. Generating an MR image from a tissue classification has traditionally been performed in MR simulation approaches using physics-based models [5]. In our approach, the MR image formation process is implicitly modeled using a convolutional neural network. Given the input MR image F and partial volumes $P(L)$, we optimize a 3D U-net with a mean squared error loss function defined by

$$\min_{\theta} \|F - \hat{F}(P(L); \theta))\|_2^2 \tag{3}$$

where θ are the network weights. U-net parameters were set similarly to the segmentation network, except the number of epochs was set to 20. Once the synthesis is learned, new atlas MR images are generated from the atlas label images within the training data. The segmentation network is updated with this new training data, and the input image is once again segmented to compute a new segmentation estimate.

3 Results

We applied the U-net segmentation network trained with MPRAGE images and CAMELION to the SPGR images of 8 held out subjects. For additional comparison, we also applied the nonlinear histogram matching (NHM) method of [17]. To evaluate consistency, the segmentation resulting from applying the original U-net segmentation network to the corresponding MPRAGE image of the subject was used as the ground truth reference. Convergence within CAMELION was empirically set to 5 iterations, which typically resulted in fewer than 5% of voxels changing labels. Note that each of the test images was processed completely independently, with the synthesis step trained using only the input image.

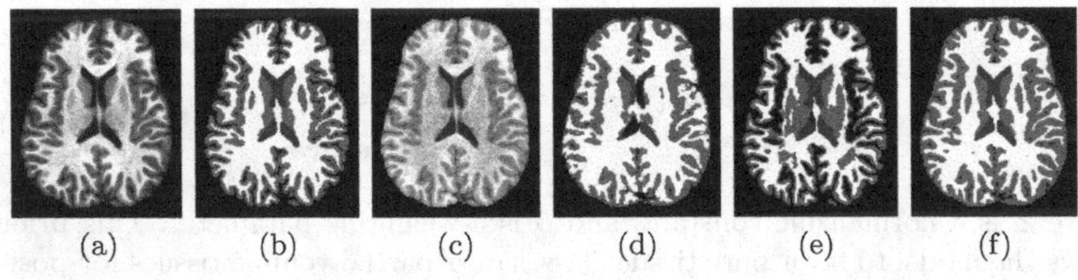

Fig. 4. Segmentation results: (a) MPRAGE image, (b) MPRAGE segmented with U-net, (c) SPGR image, (d) SPGR segmented with U-net, (e) SPGR segmented with U-net after nonlinear histogram matching, (f) SPGR segmented with CAMELION.

Figure 4 shows the results of different segmentation approaches on the SPGR data at a midventricular axial slice. Figures 4(a)–(b) show the MPRAGE image and the U-net segmentation result that serves as the reference. Because the U-net segmentation is trained with MPRAGE data acquired with the same protocol, the segmentation result is reasonably accurate. Figures 4(c)–(d) show the associated SPGR image from the same subject, as well as the segmentation when applying the same U-net. Because the SPGR contrast differs from the MPRAGE atlas images used in the training data, the white matter is overestimated and the gray matter is underestimated. Figure 4(e) shows the U-net segmentation applied to the SPGR after it has been intensity corrected using NHM. Although the cortical gray matter is better estimated in this result, the subcortical gray matter is over estimated. Figure 4(f) shows the CAMELION result applied to the SPGR images. By adapting the appearance of the atlas image to the SPGR input, the result is much more consistent with what would be obtained with an MPRAGE input. Note that there are some differences at the cortical gray matter and sulcal CSF boundaries. This is at least in part because the FreeSurfer algorithm does not explicitly segment sulcal CSF, leading to some ambiguity in the segmentation learning in these regions.

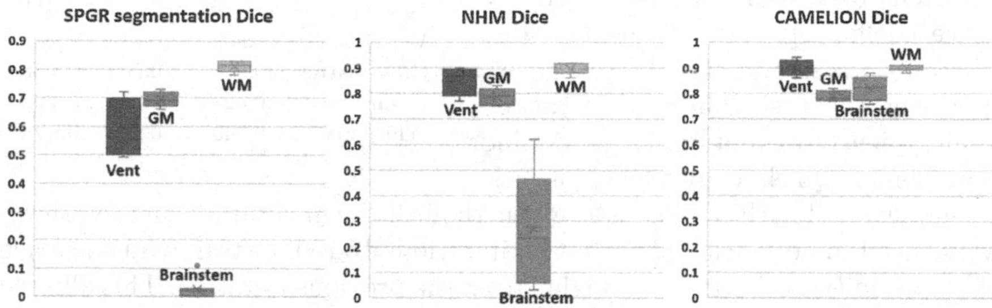

Fig. 5. Quantitative box plot comparison of Dice coefficients using three different approaches. Four tissue classes were considered: ventricles (Vent), gray matter (GM), white matter (WM), and brainstem. The center line is the median, the "x" is the mean, the box represents the first quartile, and the whiskers represent the minimum and maximum values.

Figure 5 shows quantitative results of segmenting the SPGR test images, focusing on Dice overlap in four tissue classes. Applying the U-net segmentation directly to the SPGR image yielded the lowest Dice coefficients. Adjusting the global contrast of the SPGR images with NHM generally improved results, but smaller regions such as the brainstem remain poorly segmented. Applying CAMELION further increased Dice coefficients and reduced variability of the measurements. Improvements were statistically significant relative to the standard U-net segmentation across all four tissue classes, and were significant for ventricles and brainstem relative to the NHM results ($p < 0.01$, paired t-test). Although Dice overlap provides one measure of segmentation accuracy, in research studies the primary outcome is often the total volume of a structure. Table 1 shows the Pearson correlation coefficient of volume measurements between the MPRAGE and SPGR segmentations of the same subject. Bold values show the highest correlation across methods for that tissue class. CAMELION (indicated by "CAM" in Table 1) shows improved associations between the volume measurements across the two acquisition protocols.

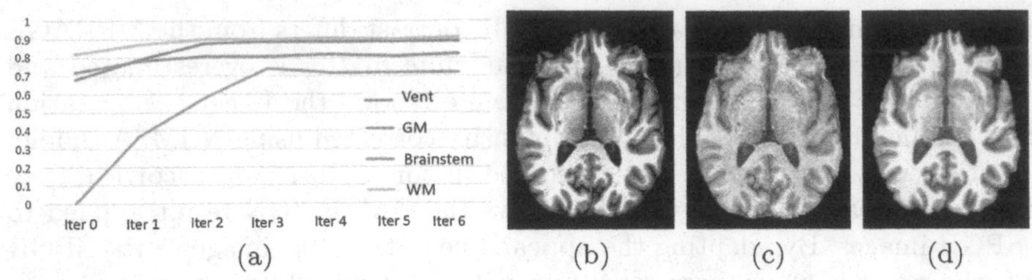

(a) (b) (c) (d)

Fig. 6. (a) CAMELION Dice coefficients across iterations, (b) Atlas MPRAGE, (b) Corresponding SPGR, (d) Synthetic SPGR.

Table 1. Volume correlations

Structure	SPGR	NHM	CAM
Ventricles	0.859	**0.999**	0.998
Gray matter	0.875	0.902	**0.938**
Brainstem	0.028	−0.258	**0.794**
White matter	0.940	0.918	**0.971**

Figure 6(a) shows the convergence of CAMELION over six iterations in terms of the Dice coefficients for a single test data set. It can be seen that the Dice increases rapidly over the first couple of iterations, and subsequently exhibits only very minor improvements. Quick convergence is relatively important for this approach, given that the segmentation and synthesis networks need to be re-trained with each iteration. To reduce training time, each network was initialized with the weights from the previous iteration. This allowed the number of epochs on the segmentation network to be reduced after the first iteration. Even so, on a Tesla V100-SXM3 with 32 GB of memory, the synthesis training required approximately 20 min per iteration, while the segmentation training required approximately 5 h for the first iteration and 2 h for subsequent iterations, leading to a total run time of approximately 14 h (5 iterations).

Figures 6(b)–(c) compare the MPRAGE and SPGR images from one of the training data subjects to the synthetic SPGR image computed after running CAMELION. The atlas images in the training data begins as a pair consisting of the MPRAGE image in Fig. 6(a) and its segmentation. With each iteration, the synthesis step transforms the MPRAGE image to gain a more SPGR-like appearance. One notable difference in the accuracy of the synthetic image compared to the original SPGR image is a lack of noise. Synthesis with U-net architectures have been shown to generally yield rather smooth images [6].

4 Discussion

CAMELION provided improved consistency in brain segmentation across two very different T1-weighted imaging protocols. The results shown here were generated essentially with off-the-shelf convolutional neural networks with very limited tuning of the architecture or network parameters. Furthermore, because there was no gold standard segmentation available on the imaging data set, we employed another segmentation algorithm to generate training data. This approach, however, had some challenges in that the atlas segmentations lacked sulcal

CSF. We believe that substantial improvements could be gained through careful optimization and the use of improved training data.

The basic framework of CAMELION bares some similarities to an autoencoder structure. In the case of CAMELION, the latent space is composed of labels or segmentations that are further transformed into partial volume estimates. Future work, will investigate the effects of the "softness" of the partial volumes, as modulated by the β parameter, on the quality of the synthesis. In addition, it may be possible to bypass the partial volume estimation entirely by directly using the probabilistic results of the segmentation network.

Although the re-training of networks within each iteration of CAMELION is computationally expensive, when working with data sets with homogeneous acquisition protocols it is possible to train the network group-wise using all available data. When a new data set with the same protocol is collected, the segmentation network does not need to be retrained and can be straightforwardly segmented without any additional iterations. Despite the fact that the data used in this work represents two homogeneously acquired data sets, each image was processed independently to demonstrate that accurate segmentation results can be achieved without training the synthesis network on multiple examples. A promising area of research will be to use CAMELION in combination with data augmentation and continual learning approaches [18] that could lead to increasingly more robust and generalizable segmentation networks.

Acknowledgements. This work was supported by a research grant from the National Multiple Sclerosis Society (RG-1907-34570), by the Department of Defense in the Center for Neuroscience and Regenerative Medicine, the intramural research program of the National Institute of Neurological Disorders and Stroke, and the intramural research program of the Clinical Center in the National Institutes of Health.

References

1. Avants, B.B., Tustison, N.J., Stauffer, M., Song, G., Wu, B., Gee, J.C.: The Insight ToolKit image registration framework. Front. Neuroinform. **8**, 44 (2014)
2. Chen, C., Dou, Q., Chen, H., Qin, J., Heng, P.A.: Unsupervised bidirectional cross-modality adaptation via deeply synergistic image and feature alignment for medical image segmentation. IEEE Trans. Med. Imaging **39**, 2494–2505 (2020)
3. Chua, A.S., et al.: Handling changes in MRI acquisition parameters in modeling whole brain lesion volume and atrophy data in multiple sclerosis subjects: comparison of linear mixed-effect models. NeuroImage Clin. **8**, 606–610 (2015)
4. Çiçek, Ö., Abdulkadir, A., Lienkamp, S.S., Brox, T., Ronneberger, O.: 3D U-Net: learning dense volumetric segmentation from sparse annotation. In: Ourselin, S., Joskowicz, L., Sabuncu, M.R., Unal, G., Wells, W. (eds.) MICCAI 2016. LNCS, vol. 9901, pp. 424–432. Springer, Cham (2016). https://doi.org/10.1007/978-3-319-46723-8_49
5. Collins, D.L., et al.: Design and construction of a realistic digital brain phantom. IEEE Trans. Med. Imaging **17**(3), 463–468 (1998)
6. Dewey, B., et al.: DeepHarmony: a deep learning approach to contrast harmonization across scanner changes. Magn. Reson. Imaging **64**, 160–170 (2019)

7. Dou, Q., Castro, D.C., Kamnitsas, K., Glocker, B.: Domain generalization via model-agnostic learning of semantic features (NeurIPS) (2019)
8. Erus, G., Doshi, J., An, Y., Verganelakis, D., Resnick, S.M., Davatzikos, C.: Longitudinally and inter-site consistent multi-atlas based parcellation of brain anatomy using harmonized atlases. Neuroimage **166**, 71–78 (2018)
9. Fischl, B.: FreeSurfer. Neuroimage **62**(2), 774–781 (2012)
10. Fortin, J.P., Sweeney, E.M., Muschelli, J., Crainiceanu, C.M., Shinohara, R.T.: Removing inter-subject technical variability in magnetic resonance imaging studies. Neuroimage **132**, 198–212 (2016)
11. George, A., Kuzniecky, R., Rusinek, H., Pardoe, H.R.: Standardized brain MRI acquisition protocols improve statistical power in multicenter quantitative morphometry studies. J. Neuroimaging **30**(1), 126–133 (2020)
12. Glocker, B., Robinson, R., Castro, D.C., Dou, Q., Konukoglu, E.: Machine learning with multi-site imaging data: an empirical study on the impact of scanner effects, pp. 1–5 (2019)
13. Iglesias, J.E., Liu, C.Y., Thompson, P.M., Tu, Z.: Robust brain extraction across datasets and comparison with publicly available methods. IEEE Trans. Med. Imaging **30**(9), 1617–1634 (2011)
14. Jog, A., Carass, A., Roy, S., Pham, D.L., Prince, J.L.: MR image synthesis by contrast learning on neighborhood ensembles. Med. Image Anal. **24**(1), 63–76 (2015)
15. Jog, A., Hoopes, A., Greve, D.N., Van Leemput, K., Fischl, B.: PSACNN: pulse sequence adaptive fast whole brain segmentation. Neuroimage **199**, 553–569 (2019)
16. Kamnitsas, K., et al.: Unsupervised domain adaptation in brain lesion segmentation with adversarial networks. In: Niethammer, M., et al. (eds.) IPMI 2017. LNCS, vol. 10265, pp. 597–609. Springer, Cham (2017). https://doi.org/10.1007/978-3-319-59050-9_47
17. Nyul, L., Udupa, J., Zhang, X.: New variants of a method of MRI scale standardization. IEEE Trans. Med. Imaging **19**(2), 143–150 (2000)
18. Parisi, G.I., Kemker, R., Part, J.L., Kanan, C., Wermter, S.: Continual lifelong learning with neural networks: a review. Neural Netw. **113**, 54–71 (2019)
19. Pham, D., Roy, S.: Alternating segmentation and simulation for contrast adaptive tissue classification. In: SPIE Medical Imaging, Houston, TX, February 10–15 2018
20. Puonti, O., Iglesias, J.E., Van Leemput, K.: Fast and sequence-adaptive whole-brain segmentation using parametric Bayesian modeling. Neuroimage **143**, 235–249 (2016)
21. Shah, M., et al.: Evaluating intensity normalization on MRIs of human brain with multiple sclerosis. Med. Image Anal. **15**(2), 267–282 (2011)
22. Shinohara, R., et al.: Volumetric analysis from a harmonized multi-site brain MRI study of a single-subject with multiple sclerosis. Am. J. Neuroradiol. **38**, 1501–1509 (2017)
23. Tustison, N.J., Avants, B.B., Cook, P.A., Zheng, Y., Egan, A., Yushkevich, P.A., Gee, J.C.: N4ITK: improved N3 bias correction. IEEE Trans. Med. Imaging **29**(6), 1310–1320 (2010)

3D Brain MRI GAN-Based Synthesis Conditioned on Partial Volume Maps

Filip Rusak[1,2(✉)], Rodrigo Santa Cruz[1,2], Pierrick Bourgeat[2], Clinton Fookes[1], Jurgen Fripp[2], Andrew Bradley[1], and Olivier Salvado[1,2]

[1] Queensland University of Technology, Brisbane, QLD 4000, Australia
[2] CSIRO, Herston, QLD 4029, Australia
filip.rusak@data61.csiro.au

Abstract. In this paper, we propose a framework for synthesising 3D brain T1-weighted (T1-w) MRI images from Partial Volume (PV) maps for the purpose of generating synthetic MRI volumes with more accurate tissue borders. Synthetic MRIs are required to enlarge and enrich very limited data sets available for training of brain segmentation and related models. In comparison to current state-of-the-art methods, our framework exploits PV-map properties in order to guide a Generative Adversarial Network (GAN) towards the generation of more accurate and realistic synthetic MRI volumes. We demonstrate that conditioning a GAN on PV-maps instead of Binary-maps results in 58.96% more accurate tissue borders in synthetic MRIs. Furthermore, our results indicate an improvement in the representation of the Deep Gray Matter region in synthetic MRI volumes. Finally, we show that fine changes introduced into PV-maps are reflected in the synthetic images, while preserving accurate tissue borders, thus enabling better control during the data synthesis of novel synthetic MRI volumes.

Keywords: Generative Adversarial Network · Partial volume maps · Synthetic MRIs · 3D image synthesis

1 Introduction

Deep Neural Networks, particularly Convolutional Neural Networks (CNNs), have demonstrated tremendous capability to perform accurate segmentation tasks when trained on large datasets [19,20]. In medical imaging, these methods are limited by the scarcity of available data. Labelling medical data is time consuming and requires a high level of expertise which is expensive. Many different CNN-based methods attempted to overcome this hurdle by mitigating the amount of data needed for their training, such as using unsupervised [3,15], weakly-supervised [8,27], semi-supervised [2,17] and self-supervised [14,21] methods. The drawback of these methods is that they are typically less accurate than supervised methods [13]. Furthermore, since the ground truth label is missing, it is more difficult to evaluate the performance of these methods [12].

N. Burgos et al. (Eds.): SASHIMI 2020, LNCS 12417, pp. 11–20, 2020.
https://doi.org/10.1007/978-3-030-59520-3_2

In contrast to aforementioned methods, data augmentation methods [5,22,23] aim to increase the number of available labelled samples needed for training of supervised methods. Data augmentation methods fall into two major trends: geometric transformation-based and GAN-based. Most geometric transformation-based augmentation methods provide limited improvement in terms of samples variety as their output highly relies on the input data.

A GAN is a data synthesis approach capable of injecting more variety into synthesised data and generating outputs less dependant of the input data, while aiming to follow the training data distribution [6]. MRI synthesis using GANs can be classified into two prominent approaches: unconditional [6,7,11] and conditional [16,22]. The main drawback of unconditional MRI synthesis approaches, in the context of supervised segmentation, is the missing segmentation labels of the newly synthesised MRIs. Another drawback of such approaches is the lack of synthesis control [16]. On the other hand, MRI synthesis approaches based on conditioning a GAN with segmentation labels, as presented in [22], keeps the brain anatomical structures intact, while segmentation labels give the ability to control the synthetic results. Nevertheless, the segmentation labels only provide an estimate of brain tissue types. Their accuracy is limited by the image resolution and consequently the segmentation accuracy may suffer from partial volume (PV) effects at the border between two tissues where a single voxel may contain multiple classes. More accurate segmentation can be represented with PV-maps as they define accurate border between two tissue classes [4], which makes them a suitable choice for conditioning GANs in the context of MRI synthesis. Conditioning GANs on PV-maps opens a pathway to generate MRIs of different appearances while retaining the same anatomical structure with fine boundary details. Having control over MRI synthesis by defining tissues with PV-maps as well as the ability to change them may be used as a powerful data synthesis approach.

In this paper, we propose a framework for synthesising 3D brain T1-weighted MRI images from PV-maps. Our proposed framework is inspired by well-known Image-to-Image conditional GAN approach described in [9]. We use PV-maps of Gray Matter (GM), White Matter (WM) and Cerebrospinal Fluid (CSF) as inputs to assist the training of the model and the generation of realistic 3D brain MRIs. We report the first attempt to synthesise realistic 3D brain MRI images from PV-maps using GANs. Furthermore, we demonstrate that changes in PV-maps reflect changes in newly generated synthetic images and show how the framework can increase the number of synthetic training images. The contributions of this paper are the following:

1) *We proposed a GAN-based framework that exploits PV-map properties to obtain synthetic MRI volumes with accurate borders between tissue classes as well as more accurate and realistic Deep Gray Matter (DGM) regions.*
2) *In the context of 3D T1-w brain MRI generation using GANs, we demonstrated that conditioning GANs on PV-maps produces better results than binary-maps. The difference is most evident in the regions of tissue borders, which is an important feature for applications such as cortical thickness estimation and segmentation.*

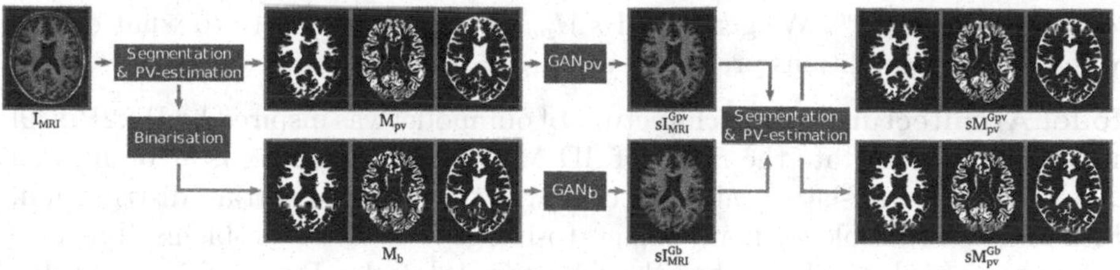

Fig. 1. Schematic representation of the experimental method.

2 Methods

Hypothesis Formulation. When it comes to T1-w brain MRI synthesis, a desirable synthetic MRI image (sI_{MRI}) is expected to respect relations between brain anatomical structures of the original MRI images (I_{MRI}). A possible method to generate such images is to condition a GAN on a particular class label to obtain results that meet the imposed condition [16]. The same mechanism may be applied to the problem of generating sI_{MRI} that keeps the brain anatomy intact. One of the simple ways is to condition a GAN for the purpose of sI_{MRI} generation with intact anatomy is to use Binary Maps (M_b) of different tissues. A M_b, in the context of 3D images, is a volume $M_b \in \{0,1\}^{w \times h \times d}$, where the value of each voxel denotes affiliation to a single class (1 indicates class affiliation). In the case of brain synthesis, a GAN can be conditioned on three classes: WM, GM and CSF; where each class is represented as a M_b. The limitation of such a class labelling method is the indivisible nature of voxel affiliation. In certain regions of an MRI, especially in the region around a tissue border, the voxel may not be of an adequate size. This limitation can be overcome by using PV-maps (M_{pv}) which, in the context of 3D images, is defined as a volume $M_{pv} \in [0,1]^{w \times h \times d}$, where the value of each voxel represents the proportion of affiliation to a single class (1 indicates 100% class affiliation). The main advantage of M_{pv} is the ability to represent partial affiliation to a certain class, which allows tissue labelling with higher precision when compared to single-class voxels.

We hypothesise that conditioning a GAN with M_{pv} instead of M_b results with better sI_{MRI}, especially at tissue interfaces. The hypothesis was evaluated by the experimental method presented in Fig. 1. The Fig. 1 shows the generation of M_{pv} from I_{MRI} by performing brain segmentation, implemented with the Expectation-maximisation (EM) algorithm [25], followed by PV-estimation implemented as in [1]. Three M_{pv}s are derived from I_{MRI}, one for each tissue-type (WM, GM and CSF). We binarise M_{pv}s by assigning each voxel to the M_{pv} with the highest partial affiliation for a particular voxel and obtain the corresponding M_b for each class. Two models were trained, GAN_{pv} on M_{pv} and GAN_b on M_b and used to generate synthetic images, sI_{MRI}^{Gpv} and sI_{MRI}^{Gb} respectively. Once the sI_{MRI} were synthesised, the reverse process was performed, where sI_{MRI}^{Gpv} and sI_{MRI}^{Gb} were segmented followed by PV-estimation in order to obtain the synthetic M_{pv} (sM_{pv}). sM_{pv} derived from sI_{MRI}^{Gb} are denoted as sM_{pv}^{Gb}, while sM_{pv} derived from sI_{MRI}^{Gpv}

are denoted as sM_{pv}^{Gpv}. We generated sM_{pv} in order to evaluate to what extent are the imposed conditions preserved in sI_{MRI}.

Model Architecture. The architecture of our model was inspired by Pix2Pix [9] and adapted to facilitate the needs of 3D MRI images. Pix2Pix is a conditional GAN capable of translating labels into images that follow a certain distribution, which makes it suitable for many image-to-image translation problems. The network is composed of a U-net-based generator [18] and a PatchGAN-based discriminator that compares image patches instead of whole images [9]. The modified architecture and its hyper-parameters are presented in Fig. 2.

We denote data of a certain distribution d_x with x, generator with G, its output $G(c_{1-3}, z)$ and discriminator with D. Moreover, we denote three condition variables with $c_{1-3}(M_b$ or M_{pv} for three tissue-types) and a noise variable with z. The objective function is defined as follows,

$$\min_G \max_D \mathbb{E}_{c_{1-3},x}\left[log\left(D\left(c_{1-3},x\right)\right)\right] + \mathbb{E}_{c_{1-3},z}\left[log\left(1 - D\left(c_{1-3}, G\left(c_{1-3}, z\right)\right)\right)\right] + \mathbb{E}_{c_{1-3},x,z}\left[\left\|x - G\left(c_{1-3}, z\right)\right\|_1\right], \tag{1}$$

where G has a goal to minimise the probability of D performing a correct binary classification task, while D aims to maximise the same. Referring to [9], we also added the L1 distance clause to the objective function as L1 tends to mitigate blurriness in the resulting images, which is needed for generation of images with accurate tissue borders. We also used the noise z in the form of dropout (activated at training and inference) across a number of layers instead of providing it as an input.

Data. For the evaluation of our training method we used a subset of 3T scans ($181 \times 218 \times 181$ voxels) from the ADNI [10,26][1] dataset. The subset contained 700 baseline subjects where only 3D T1-w volumes were used. Subjects were split into train and test sets. The train set included 500 subjects, while the 200 remaining subjects were used for the test set. All volumes were pre-processed by applying: (i) bias field correction in the brain region [24], (ii) rigid registration to the MNI-space and (iii) zero-mean normalisation with the mean value computed from the voxels in brain region of interest (ROI) only.

Training. We trained our models for 200 epochs. For the training of both models we used Adam optimiser, batch size of 1 and initial learning rate of 0.0002. After 100 epochs, we reduced the learning rate by 2×10^{-6} every epoch.

[1] Data used in the preparation of this article were obtained from the Alzheimer's Disease Neuroimaging Initiative (ADNI) database (adni.loni.usc.edu). The ADNI was launched in 2003 as a public-private partnership, led by Principal Investigator Michael W. Weiner, MD. The primary goal of ADNI has been to test whether serial magnetic resonance imaging (MRI), positron emission tomography (PET), other biological markers, and clinical and neuropsychological assessment can be combined to measure the progression of mild cognitive impairment (MCI) and early Alzheimer's disease (AD). For up-to-date information, see www.adni-info.org.

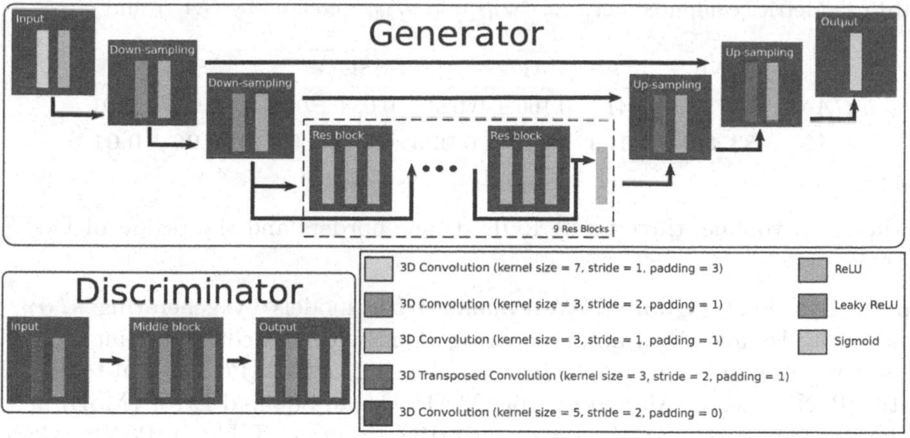

Fig. 2. Model architecture with supplementary hyper-parameter details.

3 Experiments

Our experiments were constructed to asses the benefit of using M_{pv} over M_b for the purpose of synthesising T1-w brain MRI volumes with accurate tissue-borders. Moreover, as a proof of concept for MRI synthesis, we assessed the reflection of fine changes, introduced on the M_{pv}, in sI_{MRI} and sM_{pv}. In the following experiments we evaluated the quality of sI_{MRI} and sM_{pv} on the level

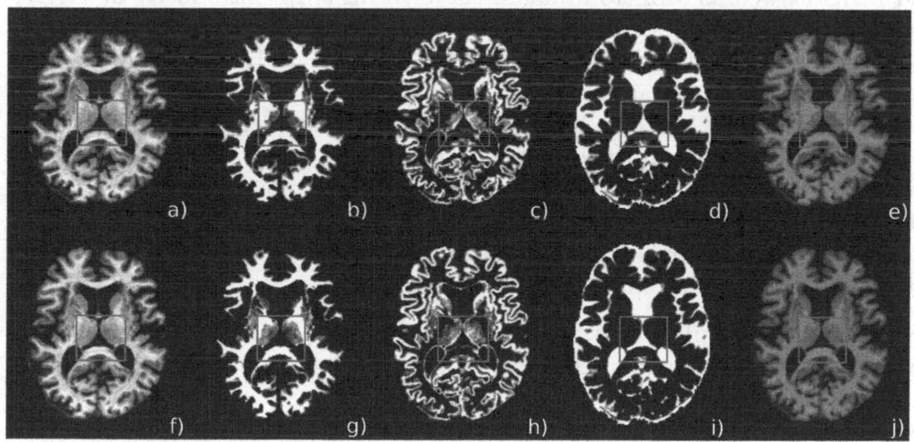

Fig. 3. Qualitative results of our framework trained on M_{pv}. Presented results show that changes introduced in the M_{pv} are reflected in the sI_{MRI}. The ground truth (a, f), two sets of M_{pv} from the same subject as well as corresponding sI_{MRI} (e, j) are presented respectively, where the region of DGM in the first case (b, c, d) is weakly defined, in comparison to the second case (g, h, i).

Table 1. Metrics computed between I_{MRI} and sI_{MRI} created by GAN_b and GAN_{pv}.

	PSNR	MAE	MSE	SSIM
GAN_b	32.777 ± 1.041	0.166 ± 0.024	0.054 ± 0.014	0.955 ± 0.01
GAN_{pv}	$\mathbf{33.449 \pm 1.103}$	$\mathbf{0.144 \pm 0.023}$	$\mathbf{0.047 \pm 0.013}$	$\mathbf{0.96 \pm 0.01}$

of the brain volume, three tissue ROIs, tissue borders and the region of DGM (Fig. 3).

Image Synthesis Quality. We evaluated our models by generating sI_{MRI} from both, M_b and M_{pv}, and comparing them with the corresponding I_{MRI}. Images were compared by employing the following metrics: Peak Signal-to-Noise Ratio (PSNR), Mean Absolute Error (MAE), Mean Squared Error (MSE) and Structural Similarity (SSIM) (see quantitative results in Table 1). PSNR, MSE and MAE were computed in the brain ROI. The dynamic range measured in the brain ROI of I_{MRI} spans between $[-0.56, 9.88]$, and was used to compute PSNR. SSIM was calculated on the whole volume, with background values set to zero as our generator generates brain sI_{MRI} without a background. Table 1 shows that GAN_{pv} produced sI_{MRI} more similar to I_{MRI} than GAN_b.

Evaluation at Tissue Level. We took a closer look and evaluated the quality of sI_{MRI} as well as the corresponding sM_{pv} in the ROI for every tissue-class (WM, GM and CSF). The GAN, segmentation and PV estimation may introduce errors in either sI_{MRI} or sM_{pv}. Therefore, we computed MAE and MSE between I_{MRI} and sI_{MRI} in order to evaluate the error introduced by GAN. We also used the Dice similarity metric (DSM) to evaluate the overlap with the ground truth and MAE as well as MSE to evaluate the error in sM_{pv} introduced by GAN, segmentation and PV estimation. Quantitative results of the error metrics for each tissue type, calculated on sI_{MRI}, are presented in Table 2. Quantitative measurements of shape and intensity error for each tissue-type computed on sM_{pv} are presented in Table 3. We concluded that less error was introduced in case of GAN_{pv}, for all three tissues. Further, sM_{pv} are more similar to the ground truth in case of GAN_{pv} where smaller shape and intensity errors were introduced. According to Table 3, CSF has a lower DSM than WM and GM for both GANs. The rational behind it is the nature of T1-w images where CSF is difficult to distinguish from the other non-brain tissues.

Evaluation of Multi-class Voxels. In this experiment, we quantitatively evaluated multi-class voxels, their position and intensity values. Quantitative

Table 2. Tissue-wise validation of sI_{MRI}. MAE and MSE are computed between I_{MRI} and sI_{MRI} inside each tissue class.

	MAE			MSE		
	WM	GM	CSF	WM	GM	CSF
GAN_b	0.03 ± 0.003	0.056 ± 0.005	0.046 ± 0.007	0.009 ± 0.001	0.016 ± 0.002	0.028 ± 0.007
GAN_{pv}	$\mathbf{0.014 \pm 0.003}$	$\mathbf{0.027 \pm 0.004}$	$\mathbf{0.032 \pm 0.007}$	$\mathbf{0.003 \pm 0.001}$	$\mathbf{0.006 \pm 0.001}$	$\mathbf{0.022 \pm 0.007}$

Table 3. Tissue-wise shape validation of sI_{MRI} and measurements of errors injected into sI_{MRI} by a *GAN* segmentation and PV estimation.

	DSM			MAE			MSE		
	WM	GM	CSF	WM	GM	CSF	WM	GM	CSF
GAN_b	0.959	0.947	0.922	0.067 ± 0.007	0.098 ± 0.007	0.114 ± 0.018	0.019 ± 0.003	0.028 ± 0.003	0.07 ± 0.018
GAN_{pv}	**0.985**	**0.981**	**0.954**	$\mathbf{0.035 \pm 0.007}$	$\mathbf{0.047 \pm 0.008}$	$\mathbf{0.082 \pm 0.017}$	$\mathbf{0.007 \pm 0.002}$	$\mathbf{0.009 \pm 0.002}$	$\mathbf{0.058 \pm 0.017}$

evaluation was performed by computing DSM between M_{pv} and sM_{pv} for evaluation of their position in sI_{MRI}, while MAE and MSE were computed to measure the intensity error between I_{MRI} and sI_{MRI}. DSM measured in sI_{MRI} generated from both *GAN*s equals the value of one, which implies the location of multi-class voxels is fully preserved in sM_{pv} for both *GAN*s. We measured MAE of 0.134±0.017 and MSE of 0.03±0.008 in the multi-class voxels of sI^b_{MRI}. In the case of sI^{pv}_{MRI}, we measured MAE of $\mathbf{0.079 \pm 0.024}$ and MSE of $\mathbf{0.01 \pm 0.007}$. We also overlaid I_{MRI} with absolute errors, computed voxel-wise, between M_{pv} and sM_{pv}, to provide more information about the localisation and severity of the errors introduced by a *GAN*, segmentation and PV estimation (see Fig. 4). We found that most of the errors happen at tissue boundaries and observed errors of higher value in case of GAN_b. This result illustrates the benefit of using M_{pv} over M_b for the purpose of preserving well defined tissue borders in sI_{MRI}.

According to the presented quantitative results, we obtained **58.96%** smaller MAE and **33.33%** smaller MSE in multi-class voxels of sI^{pv}_{MRI} comparing to sI^b_{MRI}. The presented results support the illustration of absolute errors and strongly suggest that tissue-borders are preserved with higher accurately in sI_{MRI} generated by GAN_{pv} opposed to GAN_b.

Evaluation of Deep Gray Matter. The region of DGM contains voxels that belong to WM, GM or to both classes. The border between WM and DGM is vaguely defined and hard to segment. Furthermore, in the context of MRI synthesis, a loosely defined or flawed border between WM and DGM makes it easy to distinguish between I_{MRI} and sI_{MRI}. We evaluated the performance

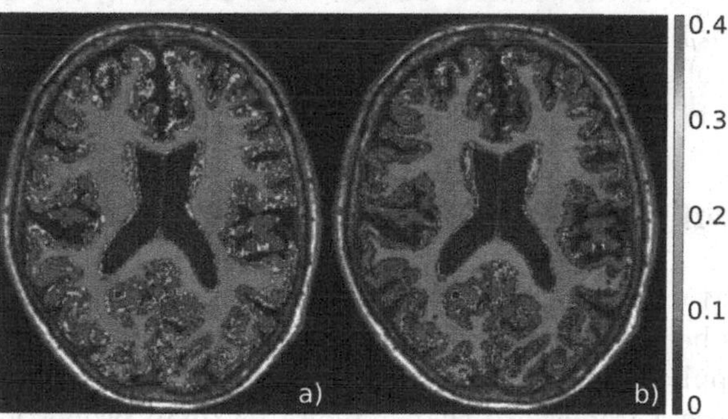

Fig. 4. Location and severity of errors injected into M_{pv} by *GAN*, segmentation and PV estimation. Absolute errors between M_{pv} and sM^{Gb}_{pv} as well as M_{pv} and sM^{Gpv}_{pv} are shown in (a) and (b), respectively.

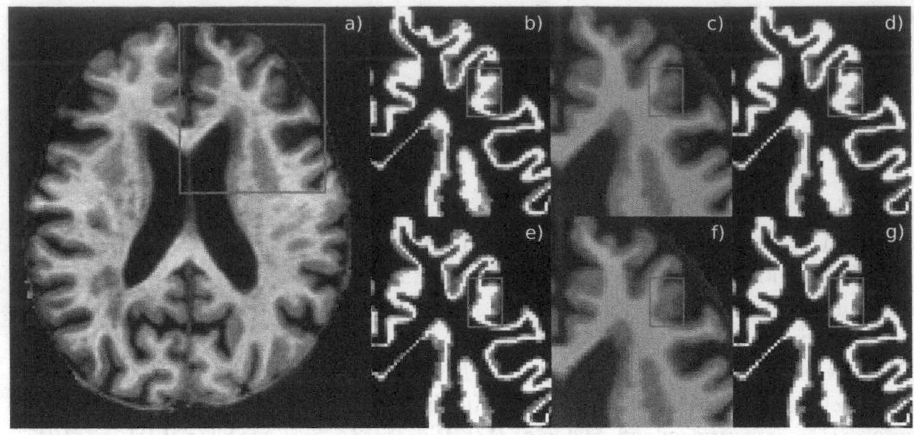

Fig. 5. Introducing fine changes. We introduced a small change into M_{pv} (b) derived from I_{MRI} (a) which is shown in (e). We generated sI_{MRI} from the original and modified M_{pv} shown in (c) and (f). The M_{pv} were then derived from sI_{MRI} in order to verify if the introduced changes are preserved in sI_{MRI}, shown in (d, g).

of both models in the region of DGM. Quantitative analysis was performed on sI_{MRI} by computing MAE and MSE to measure the error injected by a GAN. In the DGM region of sI^{b}_{MRI} we measured MAE of 0.129±0.021 and MSE of 0.029± 0.01. Yet, in the same ROI of sI^{pv}_{MRI} we measured MAE of $\mathbf{0.108 \pm 0.024}$ and MSE of $\mathbf{0.022 \pm 0.008}$. This indicates that the DGM region is more accurately represented in sI_{MRI} generated by the GAN_{pv} when compared to GAN_b.

Introduction of Fine Changes on Pv-Map Level. The outcomes of this experiment stand for a proof of concept that brain MRI synthesis may be controlled by changing M_{pv}, as the changes are reflected in the sI_{MRI}, while the model still preserves accurate tissue borders. To validate stability, we assessed the ability of the model to preserve fine changes (in this case seven voxels only) in M_{pv} by verifying if the changes are reflected in sI_{MRI}. Both the changed and unchanged M_{pv} were used to generate sI_{MRI}, which were further used to derive sM_{pv}. We obtained the introduced changes in sI_{MRI} and sM_{pv} as shown in Fig. 5.

4 Conclusion

In this work, we tackle the problem of synthesising 3D brain T1-w MRIs with accurate borders between tissues. This is an important feature in the context of medical image applications related to cortical thickness estimation and segmentation. We propose a framework that exploits PV-map properties and demonstrate that it performs better when it comes to synthetic MRI generation with accurate tissue borders compared to binary-map-based alternative. Moreover, we show that even fine changes introduced on PV-maps are reflected in synthetic images. This implies the possibility of using the framework as a data augmentation mechanism and it will be further explored in our future work.

References

1. Acosta, O., et al.: Automated voxel-based 3D cortical thickness measurement in a combined Lagrangian-Eulerian PDE approach using partial volume maps. Med. Image Anal. **13**(5), 730–743 (2009)
2. Bai, W., et al.: Semi-supervised learning for network-based cardiac MR image segmentation. In: Descoteaux, M., Maier-Hein, L., Franz, A., Jannin, P., Collins, D.L., Duchesne, S. (eds.) MICCAI 2017. LNCS, vol. 10434, pp. 253–260. Springer, Cham (2017). https://doi.org/10.1007/978-3-319-66185-8_29
3. Balakrishnan, G., Zhao, A., Sabuncu, M.R., Guttag, J., Dalca, A.V.: An unsupervised learning model for deformable medical image registration. In: Proceedings of the IEEE Conference on Computer Vision and Pattern Recognition, pp. 9252–9260 (2018)
4. Ballester, M.Á.G., Zisserman, A.P., Brady, M.: Estimation of the partial volume effect in MRI. Med. Image Anal. **6**(4), 389–405 (2002)
5. Frid-Adar, M., Klang, E., Amitai, M., Goldberger, J., Greenspan, H.: Synthetic data augmentation using GAN for improved liver lesion classification. In: 2018 IEEE 15th International Symposium on Biomedical Imaging (ISBI 2018), pp. 289–293. IEEE (2018)
6. Goodfellow, I., et al.: Generative adversarial nets. In: Advances in Neural Information Processing Systems, pp. 2672–2680 (2014)
7. Han, C., et al.: Combining noise-to-image and image-to-image GANs: Brain MR image augmentation for tumor detection. IEEE Access **7**, 156966–156977 (2019)
8. Hu, Y., et al.: Label-driven weakly-supervised learning for multimodal deformable image registration. In: 2018 IEEE 15th International Symposium on Biomedical Imaging (ISBI 2018), pp. 1070–1074. IEEE (2018)
9. Isola, P., Zhu, J.Y., Zhou, T., Efros, A.A.: Image-to-image translation with conditional adversarial networks. In: Proceedings of the IEEE Conference on Computer Vision and Pattern Recognition, pp. 1125–1134 (2017)
10. Jack Jr., C.R., et al.: The Alzheimer's disease neuroimaging initiative (ADNI): MRI methods. J. Magn. Reson. Imaging Off. J. Int. Soc. Magn. Reson. Med. **27**(4), 685–691 (2008)
11. Kwon, G., Han, C., Kim, D.: Generation of 3D brain MRI using auto-encoding generative adversarial networks. In: Shen, D., et al. (eds.) MICCAI 2019. LNCS, vol. 11766, pp. 118–126. Springer, Cham (2019). https://doi.org/10.1007/978-3-030-32248-9_14
12. Lee, J.G., et al.: Deep learning in medical imaging: general overview. Korean J. Radiol. **18**(4), 570–584 (2017)
13. Lenchik, L., et al.: Automated segmentation of tissues using CT and MRI: a systematic review. Acad. Radiol. **26**(12), 1695–1706 (2019)
14. Li, H., Fan, Y.: Non-rigid image registration using self-supervised fully convolutional networks without training data. In: 2018 IEEE 15th International Symposium on Biomedical Imaging (ISBI 2018), pp. 1075–1078. IEEE (2018)
15. Mahjourian, R., Wicke, M., Angelova, A.: Unsupervised learning of depth and egomotion from monocular video using 3D geometric constraints. In: Proceedings of the IEEE Conference on Computer Vision and Pattern Recognition, pp. 5667–5675 (2018)

16. Mirza, M., Osindero, S.: Conditional generative adversarial nets. arXiv preprint arXiv:1411.1784 (2014)
17. Pombo, G., Gray, R., Varsavsky, T., Ashburner, J., Nachev, P.: Bayesian volumetric autoregressive generative models for better semisupervised learning. In: Shen, D., et al. (eds.) MICCAI 2019. LNCS, vol. 11767, pp. 429–437. Springer, Cham (2019). https://doi.org/10.1007/978-3-030-32251-9_47
18. Ronneberger, O., Fischer, P., Brox, T.: U-Net: convolutional networks for biomedical image segmentation. In: Navab, N., Hornegger, J., Wells, W.M., Frangi, A.F. (eds.) MICCAI 2015. LNCS, vol. 9351, pp. 234–241. Springer, Cham (2015). https://doi.org/10.1007/978-3-319-24574-4_28
19. Ros, G., Sellart, L., Materzynska, J., Vazquez, D., Lopez, A.M.: The SYNTHIA dataset: a large collection of synthetic images for semantic segmentation of urban scenes. In: Proceedings of the IEEE Conference on Computer Vision and Pattern Recognition, pp. 3234–3243 (2016)
20. Roy, A.G., Conjeti, S., Navab, N., Wachinger, C., Initiative, A.D.N., et al.: QuickNAT: a fully convolutional network for quick and accurate segmentation of neuroanatomy. NeuroImage **186**, 713–727 (2019)
21. Santa Cruz, R., Fernando, B., Cherian, A., Gould, S.: DeepPermNet: visual permutation learning. In: Proceedings of the IEEE Conference on Computer Vision and Pattern Recognition, pp. 3949–3957 (2017)
22. Shin, H.-C., et al.: Medical image synthesis for data augmentation and anonymization using generative adversarial networks. In: Gooya, A., Goksel, O., Oguz, I., Burgos, N. (eds.) SASHIMI 2018. LNCS, vol. 11037, pp. 1–11. Springer, Cham (2018). https://doi.org/10.1007/978-3-030-00536-8_1
23. Taylor, L., Nitschke, G.: Improving deep learning using generic data augmentation. arXiv preprint arXiv:1708.06020 (2017)
24. Van Leemput, K., Maes, F., Vandermeulen, D., Suetens, P.: Automated model-based bias field correction of MR images of the brain. IEEE Trans. Med. Imaging **18**(10), 885–896 (1999)
25. Van Leemput, K., Maes, F., Vandermeulen, D., Suetens, P.: Automated model-based tissue classification of MR images of the brain. IEEE Trans. Med. Imaging **18**(10), 897–908 (1999)
26. Weiner, M.W., et al.: The Alzheimer's disease neuroimaging initiative 3: continued innovation for clinical trial improvement. Alzheimer's Dement. **13**(5), 561–571 (2017)
27. Xu, Y., Zhu, J.Y., Eric, I., Chang, C., Lai, M., Tu, Z.: Weakly supervised histopathology cancer image segmentation and classification. Med. Image Anal. **18**(3), 591–604 (2014)

Synthesizing Realistic Brain MR Images with Noise Control

Lianrui Zuo[1,3(✉)], Blake E. Dewey[1,4], Aaron Carass[1], Yufan He[1], Muhan Shao[1], Jacob C. Reinhold[1], and Jerry L. Prince[1,2]

[1] Department of Electrical and Computer Engineering, The Johns Hopkins University, Baltimore, MD 21218, USA
lr_zuo@jhu.edu

[2] Department of Computer Science, The Johns Hopkins University, Baltimore, MD 21218, USA

[3] Laboratory of Behavioral Neuroscience, National Institute on Aging, National Institute of Health, Baltimore, MD 20892, USA

[4] Kirby Center for Functional Brain Imaging Research, Kennedy Krieger Institute, Baltimore, MD 21205, USA

Abstract. Image synthesis in magnetic resonance (MR) imaging has been an active area of research for more than ten years. MR image synthesis can be used to create images that were not acquired or replace images that are corrupted by artifacts, which can be of great benefit in automatic image analysis. Although synthetic images have been used with success in many applications, it is quite often true that they do not look like real images. In practice, an expert can usually distinguish synthetic images from real ones. Generative adversarial networks (GANs) have significantly improved the realism of synthetic images. However, we argue that further improvements can be made through the introduction of noise in the synthesis process, which better models the actual imaging process. Accordingly, we propose a novel approach that incorporates randomness into the model in order to better approximate the distribution of real MR images. Results show that the proposed method has comparable accuracy with the state-of-the-art approaches as measured by multiple similarity measurements while also being able to control the noise level in synthetic images. To further demonstrate the superiority of this model, we present results from a human observer study on synthetic images, which shows that our results capture the essential features of real MR images.

Keywords: Deep learning · MRI · Randomness · Synthesis · Noise control

1 Introduction

In clinical studies, MR images of the same anatomy with different contrasts are often acquired in order to reveal different biological information. For example, T1-weighted (T1-w) images may be acquired to show a balanced contrast between various tissues, T2-weighted (T2-w) images may be acquired to

N. Burgos et al. (Eds.): SASHIMI 2020, LNCS 12417, pp. 21–31, 2020.
https://doi.org/10.1007/978-3-030-59520-3_3

show contrast between fluid and tissues, and fluid-attenuated inversion recovery (FLAIR) images are often used to detect white matter lesions [4]. However, it is not always possible to acquire a complete set of contrasts in practice, since MR acquisition is relatively time consuming and may result in patient discomfort. Image artifacts can severely affect image quality, thus making the acquired image less informative. This is especially undesirable and more likely in longitudinal studies when several scans have poor image quality and cannot be used in downstream analysis. This often results in the exclusion of all the scans of a patient, even though many of the excluded images are good. In these non-ideal scenarios, image synthesis can be used to generate new images with missing contrasts from existing contrast images with the same anatomy.

An observed MR image is mainly determined by the anatomy being imaged and the imaging parameters that are used in image acquisition [4]. When the anatomy is fixed, choosing different imaging parameters will produce images with different contrasts. Although the observable anatomical information may vary slightly with different imaging parameters, it is often assumed that MR images of the same anatomy but different contrast (approximately) share the same underlying anatomical information. Based on this assumption, many MR synthesis approaches have been proposed in recent years (cf. [2,6,9]). The main idea is to learn an intensity transformation $T(\cdot)$ that maps intensity x_s from source image to target intensity x_t, while preserving the anatomy, i.e., $\hat{x}_t = T(x_s) \approx x_t$. The learned transformation $T(\cdot)$ is typically a deterministic function, which means that for the same input x_s, the exact same output $\hat{x}_t$ will be obtained (if we ignore spatial encoding). Although downstream tasks may be facilitated by image synthesis, the images often lack realism. Synthetic images are typically somewhat blurry and too "clean" (e.g., almost noise free), and an expert can usually distinguish synthetic images from real ones. The absence of realism can be understood as a distribution mismatch between real and synthetic MR images. If the goal of image synthesis is to replace real data, then additional effects such as denoising and spatial smoothing should not be included in the image synthesis process. This may be particularly relevant when synthesis is being used for data augmentation in training a deep network for an image analysis task.

In this paper, we argue that the distribution mismatch between real and synthetic MR images comes from the way that synthetic images are generated; most synthesis methods learn a deterministic function while real image acquisition involves random effects such as acquisition noise in MR images. The existence of randomness in MR imaging has long been ignored by the image synthesis community. Although generative adversarial networks (GANs) [10] have been applied to solve the distribution mismatch problem in medical image synthesis [2,13,17], their intensity transformation is still deterministic, which is not sufficient to model random effects in real MR images. Therefore, many GAN-based methods are only able to "memorize" and pass through the exact *noise samples* from a source image to target and synthesize pseudo noisy images, but not able to learn higher statistics to actually generate such noise as in real image acquisition. To address the issue, we propose a novel approach that explicitly approximates the distribution of real MR images with randomness incorporated. The contributions of the paper can be summarized as:

- We propose an approach to alleviate the distribution mismatch problem in synthesis models. In contrast to GANs, synthetic images are generated by drawing samples from the estimated distribution in a nondeterministic way.
- We are able to control the level of randomness (e.g., noise level) in synthetic images, which could potentially be used in MR data augmentation or other MR validation experiments.

2 Method

2.1 Incorporating Randomness in Synthesis

The training goal in image synthesis is to find an optimal mapping $T_{\text{syn}}(\cdot)$ such that the synthesis error $\mathcal{L}$, given source intensity values x_s and the target intensity values x_t, is minimized, i.e., $T_{\text{syn}}(\cdot) = \arg\min_{T(\cdot)} \mathcal{L}\left(T(x_s), x_t\right)$. Once a synthesis model is trained, an estimate of the target intensity can be calculated using $\hat{x}_t = T_{\text{syn}}(x_s)$. However, randomness is inevitable in MR imaging, which means that the observed image intensity will be different from its underlying true value, in general. In other words, when we acquire an MR image of an anatomy with selected imaging parameters, we are essentially drawing a sample from the distribution $p(x|\dot{x})$ at each voxel, where x and $\dot{x}$ are the observed voxel intensity and the underlying true intensity, respectively. In MR image synthesis, we have target image intensity $x_t \sim p(x_t|\dot{x}_t)$. An example of $p(x|\dot{x})$ can be noisy MR data, where the observed intensity follows a Rician distribution [3], i.e., $p\left(x|\dot{x};\sigma\right) = \frac{x}{\sigma^2} e^{-(x^2+\dot{x}^2)/2\sigma^2} I_0\left(\frac{\dot{x}x}{\sigma^2}\right)$, where σ controls the noise level and I_0 is the zeroth order Bessel function. Although acquisition noise can be a main source of randomness in MR data and modeled by a Rician distribution, there are many other random effects that follow very complex distributions and are impossible to model in closed-form equations. Considering the inherent randomness in MR images, we solve the following regularized optimization problem in this work,

$$T_{\text{syn}}(\cdot) = \arg\min_{T(\cdot)} \mathcal{L}\Big(T(x_s, \epsilon), x_t\Big) + \lambda \mathcal{D}\Big(p(T(x_s, \epsilon)), p(x_t)\Big), \tag{1}$$

where $\epsilon \sim \mathcal{N}(0, I)$ is a random vector that introduces additive feature noise in both training and testing, and $\mathcal{D}$ represents the dissimilarity between two distributions. In this work, we use the l_1 loss and perceptual loss [7] as our synthesis error function $\mathcal{L}$, and an adversarial loss [10] to penalize the dissimilarity between two distributions. The implications of both Eq. 1 and the introduced ϵ are discussed in Sect. 2.2.

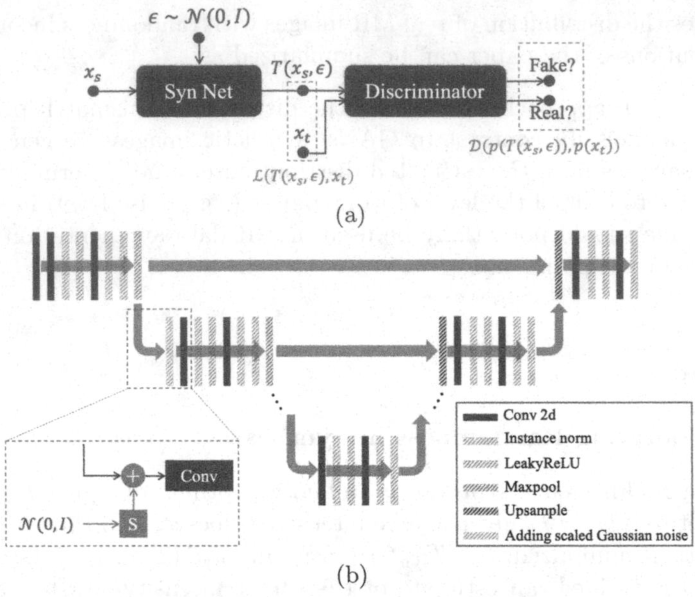

Fig. 1. (a) The framework of our proposed method during training, and (b) the structure of Syn-Net, which has a U-net like structure with scaled Gaussian noise after the maxpooling layer. The block S represents learnable parameter $\boldsymbol{s}$. It controls the standard deviation of Gaussian noise, where $\boldsymbol{s} = [s_1, s_2, \ldots, s_C]$ and C is channel number of the input feature map.

2.2 Network Structure

For simplicity, we adopt a similar U-net structure as in [12] as our main synthesis network (Syn-Net), which consists of four levels (four downsampling layers and four upsampling layers). The numbers of feature channels from the image level to the very bottom level of the Syn-Net are 1, 8, 16, 32, 64, and 128, respectively. All the convolution kernels are 3-by-3 with stride 1. We use bilinear interpolation followed by a convolution layer as our upsampling block. After each upsampling block, feature maps are concatenated and sent to the consecutive layers. The design of our discriminator is four convolution layers followed by three fully connected layers. The synthesis network takes a T2-w image as input, and its goal is to generate a synthetic T1-w image. As shown in Fig. 1, after the first downsampling layer (level #1 of the Syn-Net), each feature map has scaled Gaussian noise added per-channel, and it is then sent to successive layers. Similar to the style-GAN [8], scalars that control the standard deviation of the Gaussian noise are learnable parameters. In contrast to the style-GAN, which has scaled Gaussian noise after *each* convolution layer, we only introduce one level of randomness. The reason for this difference comes from the different purposes of the two approaches. The style-GAN aims at doing style transfer; thus, adding multi-level randomness helps the network generate images with diversity at each feature level. On the other hand, since our task is image synthesis, pixel-to-pixel accuracy—i.e., geometric accuracy—is an important constraint. Therefore, only a limited amount of randomness is needed in this work such that some degree of acquisition noise and variations in tissue intensities are introduced, while the geometry of the underlying anatomy is untouched. After passing an input image to our synthesis network, we use a discriminator to judge whether the synthetic image is a real or fake MR image, and the loss is backpropagated to the synthesis network accordingly. Note that there are other possibilities to introducing randomness to the synthesis network. One possible solution is to concatenate Gaussian noise to the input image. We will show comparison results of the proposed method and other potential solutions in Sect. 3.1.

There are two main aspects that distinguish the proposed method from existing MR image synthesis approaches. First, GANs are designed to learn a function of random variables, such that the transformed random variable approximately follows the target distribution. In our case, by adding randomness and introducing a discriminator, the synthesis network actually learns a transformation $T_{\text{syn}}(x_s, \epsilon)$ that maps its input variables $x_s \sim p(x_s)$ and $\epsilon \sim \mathcal{N}(0, I)$ to match the target distribution $p(x_t)$, i.e., $p(T_{\text{syn}}(x_s, \epsilon)) \approx p(x_t)$. This motivates the second term of Eq. 1. Second, with the introduced randomness ϵ, it becomes easier for the synthesis network to learn a probability mapping. As pointed by [8], without injecting randomness, the network has to sacrifice some of its capacity to generate pseudo-randomness. Moreover, once the network is trained, a synthetic image is generated by sampling $p(T_{\text{syn}}(x_s, \epsilon))$, which approximates the way that real images are acquired. By minimizing the synthesis error function (first term) in Eq. 1, the sampling will only change randomness like noise in synthetic images, but not anatomy. Once the whole network is trained, we can manipulate the standard deviation of ϵ to synthesize MR images with different noise levels. It is worth mentioning that although the network in [2] has a similar structure, it lacks randomness; thus, it still learns a deterministic function $T_{\text{syn}}(\cdot)$, which is fundamentally different from the proposed method.

3 Experiments

Experiments were performed on the publicly available IXI dataset [1]. The images used in the experiments were acquired from Guys Hospital by a Philips 1.5T scanner. Each subject has both T1-w (TE = 4.603 ms, TR = 9.8 ms) and T2-w (TE = 100 ms, TR = 8.2 s) MR images. Before synthesis, each subject was preprocessed with N4 inhomogeneity correction [15], white matter peak normalization [11], and then co-registered to the MNI space with 1 mm^3 isotropic resolution. 162 subjects were used in training, 59 subjects in validation and 97 subjects for testing. For each subject, we evenly selected 21 axial cross sections from the center 60 mm of the image with spacing between each slice around 3 mm.

Table 1. Synthesis performance of different network settings under multiple measurements. SSIM and PSNR are presented in the format of "mean ± standard deviation". Specifically, we studied the effects of whether randomness is introduced and whether a discriminator is applied. Rows a–d: synthesis results with $\epsilon \sim \mathcal{N}(0, I)$ in both training and testing. Rows e–g: synthesis results with increased noise insertion, i.e., $\epsilon \sim \mathcal{N}(0, 10I)$, during testing.

Index	ϵ	Random	Discri.	SSIM	PSNR	FID
a	$\mathcal{N}(0, I)$	×	×	$\mathbf{0.956 \pm 0.017}$	34.79 ± 1.79	11.958
b		×	√	0.952 ± 0.020	34.36 ± 2.01	5.725
c		√	×	0.956 ± 0.018	$\mathbf{34.80 \pm 1.78}$	12.425
d		√	√	0.953 ± 0.020	34.35 ± 1.90	**4.902**
e	$\mathcal{N}(0, 10I)$	√	×	0.932 ± 0.019	33.00 ± 1.60	21.653
f		√ (concat.)	√	0.911 ± 0.024	32.42 ± 1.69	25.757
g		√	√	$\mathbf{0.937 \pm 0.020}$	$\mathbf{33.22 \pm 1.69}$	**5.551**

3.1 Ablation Study

We performed two ablation studies to show the synthesis performance with different network settings. The first study aims to show the effect of introducing noise to a Syn-Net/Discriminator setting. The second study compares the results of introducing noise at different locations of the synthesis model.

Effects of Introducing Random Noise. In the first study, all the settings share the same synthesis error function $\mathcal{L}$ (i.e., l_1 and perceptual loss), and the differences come from whether we inject randomness or apply a discriminator. We use two types of similarity measurements as our evaluation criteria: intensity based and distribution based. The intensity based similarity measurements include the structural similarity index (SSIM) [16] and the peak signal to noise ratio (PSNR). For distribution based metric, we use the Fréchet inception distance (FID) [5], which is a metric on distributions, as it compares statistics of feature maps between generated samples and real samples. The FID is often used to quantify performance of GAN-based algorithms [8], and it is calculated using the pretrained InceptionV3 net [14]. A lower FID score indicates a higher similarity between two distributions.

Table 1 provides quantitative results with different network settings. We adopt a similar network structure as in [2] (Table 1 row b) as our baseline method since it was reported to have the state-of-the-art performance in intensity based metrics compared with other work. To justify the proposed method is able to synthesize images with different noise levels, we also considered other network settings. For default settings where $\epsilon \sim \mathcal{N}(0, I)$, several observations can be made from the table. First, all four methods have similar SSIM and PSNR score. Notice that Table 1 row d slightly decreases the SSIM and PSNR. We argue that the decreasing results from the fact that the proposed method

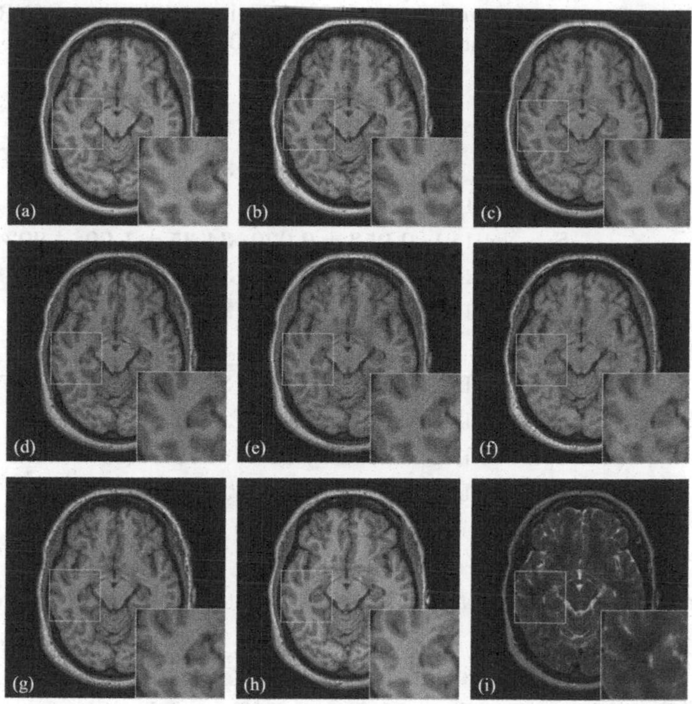

Fig. 2. Visual comparison (electronic zoom-in recommended). (a)–(g) correspond with the network settings in Table 1 rows a–g, respectively. (h) The truth image. (i) The source image.

encourages noise in synthetic images. Even though a synthetic image may have similar *noise level* as the corresponding truth image, it is not possible to generate an image with the same *noise sample* as the truth image. Second, the lower FID score in Table 1 rows b and d indicate that introducing a discriminator reduces the distribution mismatch between synthetic images and truth images. We show in Sect. 3.2 that the FID score has association with human perception. Figure 2 provides a visual comparison of the proposed method with other methods.

In our next experiment, we increased the noise levels of synthetic images by increasing the standard deviation of ϵ tenfold, i.e., $\epsilon \sim \mathcal{N}(0, 10I)$. We also compared the proposed method with other methods that could synthesize noisy MR images. Specifically, we compared two candidate methods: using the same network structure as the proposed method but without a discriminator (Table 1 row e), and concatenating the input image with Gaussian noise (Table 1 row f). As shown in Figs. 2(e) and (f), both candidate methods generate synthetic images with obvious artifacts which do not exist in real MR images, while the proposed method (Fig. 2(g)) is able to synthesize realistic MR images with multiple noise levels. Notice that in the high noise group of Table 1, the proposed method has the best performance in all three measurements, and importantly,

Table 2. Effects of introducing noise at different locations. SSIM and PSNR are presented in the format of "mean±standard deviation". Bold numbers indicate the best performance within each group.

	Location	SSIM	PSNR	FID
$\epsilon \sim \mathcal{N}(0, I)$	Image	0.951 ± 0.021	34.31 ± 1.68	7.768
	Syn-Net Lv0	0.949 ± 0.020	33.79 ± 1.67	7.760
	Syn-Net Lv1	$\mathbf{0.953 \pm 0.020}$	$\mathbf{34.35 \pm 1.90}$	**4.902**
	Syn-Net Lv2	0.952 ± 0.020	33.95 ± 1.80	4.971
	Syn-Net Lv3	$\mathbf{0.953 \pm 0.020}$	33.49 ± 1.91	5.051
$\epsilon \sim \mathcal{N}(0, 10I)$	Image	0.911 ± 0.024	32.42 ± 1.69	25.757
	Syn-Net Lv0	0.921 ± 0.021	$\mathbf{33.23 \pm 1.67}$	10.887
	Syn-Net Lv1	$\mathbf{0.937 \pm 0.020}$	33.22 ± 1.69	**5.551**
	Syn-Net Lv2	0.937 ± 0.022	33.14 ± 1.69	5.562
	Syn-Net Lv3	0.930 ± 0.021	33.05 ± 1.88	5.779

the SSIM and PSNR still remain high, which indicates that increasing noise level does not change anatomy.

Effects of Introducing Random Noise at Different Levels. In the second ablation study, we keep the network setting the same as Table 1 row d, and compare the effects of introducing randomness at different locations. Specifically, we compare the effects of introducing noise in the image space by concatenation and different levels of the Syn-Net. Results are shown in Table 2. According to our experiments, introducing noise at a relatively deeper level (i.e., level #1 or deeper) provides better synthesis performance compared with the results of shallower noise injection, especially in the noisy scenario. We hypothesize that due to the skip connections of the U-net, noise should be injected into a deeper level to utilize richer network capacity. Another observation is that there is no statistically significant performance difference from level #1 to level #3 (under paired t-tests). Since there is a synthesis error term in Eq. 1, which limits the allowable randomness to be added to the synthetic images, the proposed model is not sensitive to the level where the randomness is introduced.

3.2 Observer Study

To show that the proposed method generates more realistic MR images as compared to other methods, we conducted an observer study, as follows. A human rater with MR background is shown one MR image at a time, and the rater is asked to make a binary decision: is the image real or fake? Among our five human raters, the person with the minimum expertise has 2–3 years of experience in MR research, and the maximum has more than 15 years of experience. The average experience is 6 years. There are in total 100 MR images coming

Table 3. Misclassification rate of the observer test. The top five rows correspond to the Index (a) to (d), and (g) in Table 1. Numbers in smaller font represent the number of misclassified examples in the "inferior/middle/superior" regions of the brain. Bold numbers represent the highest misclassification rate of that rater on synthetic images.

Rand.	Discri.	Rater 1	Rater 2	Rater 3	Rater 4	Rater 5
×	×	0/15 0/0/0	1/15 0/0/1	0/15 0/0/0	2/15 2/0/0	1/15 0/1/0
×	√	3/15 3/0/0	**9/15** 5/1/3	8/15 5/1/2	5/15 3/0/2	6/15 4/0/2
√	×	0/15 0/0/0	4/15 2/1/1	4/15 2/1/1	0/15 0/0/0	2/15 1/0/1
√	√	6/15 4/1/1	6/15 3/2/1	9/15 5/3/1	5/15 4/0/1	**10/15** 5/1/4
√ (noisy)	√	**10/15** 5/4/1	3/15 1/1/1	**15/15** 5/5/5	**6/15** 4/1/1	7/15 3/1/3
– (real)	–	6/25 1/3/2	9/25 2/6/1	3/25 0/3/0	11/25 0/6/5	2/25 1/1/0

from six different sources: the four settings from Table 1 rows a–d (15 images each), proposed method with higher noise (Table 1 row g, 15 images), and 25 real MR images. The 100 images were evenly selected to cover the brain regions from the inferior lobe to superior lobe. We excluded the two candidate methods in Figs. 2(e) and (f) from the observer study because of the obvious artifacts.

Table 3 shows the misclassification rate of five human raters. Specifically, for synthetic images, misclassification means that a fake image is rated as real, and vice versa. We see that the highest misclassification rate of all five raters occurs in the methods that have a discriminator. Three out of five raters misclassified the most synthetic images as real on the noisy settings of the proposed method. This also indicates a strong association between the FID score and the misclassification rate. We also see that the highest misclassification rate occurs around the inferior region of the brain. This could be because MR images are naturally noisy in this region. On the other hand, for real images, the middle region, where the lateral ventricles are the most prominent, the misclassification rate is highest.

4 Discussion and Conclusion

In this paper, we proposed an approach to incorporate randomness into a synthesis network that can generate synthetic MR images with different noise levels. An observer study suggests that images synthesized by the proposed method are less distinguishable from real MR images than competing methods. The method could be used for conventional applications such as image imputation and artifact reduction as well as data augmentation where different levels of noise are important to explore. Future work may include disentangling anatomy noise and acquisition noise during synthesis.

Acknowledgments. This research was supported by the Intramural Research Program of the NIH, National Institute on Aging and by the TREAT-MS study funded by the Patient-Centered Outcomes Research Institute (PCORI).

References

1. IXI Brain Development Dataset. https://brain-development.org/ixi-dataset/. Accessed 10 Dec 2019
2. Dar, S.U., Yurt, M., Karacan, L., Erdem, A., Erdem, E., Çukur, T.: Image synthesis in multi-contrast MRI with conditional generative adversarial networks. IEEE Trans. Med. Imaging **38**(10), 2375–2388 (2019)
3. Gudbjartsson, H., Patz, S.: The Rician distribution of noisy MRI data. Magn. Reson. Med. **34**(6), 910–914 (1995)
4. Haacke, E.M., Brown, R.W., Thompson, M.R., Venkatesan, R.: Magnetic Resonance Imaging: Physical Principles and Sequence Design . Wiley, Hoboken (1999)
5. Heusel, M., Ramsauer, H., Unterthiner, T., Nessler, B., Hochreiter, S.: GANs trained by a two time-scale update rule converge to a local Nash equilibrium. In: Advances in Neural Information Processing Systems, pp. 6626–6637 (2017)
6. Jog, A., Carass, A., Roy, S., Pham, D.L., Prince, J.L.: Random forest regression for magnetic resonance image synthesis. Med. Image Anal. **35**, 475–488 (2017)
7. Johnson, J., Alahi, A., Fei-Fei, L.: Perceptual losses for real-time style transfer and super-resolution. In: Leibe, B., Matas, J., Sebe, N., Welling, M. (eds.) ECCV 2016. LNCS, vol. 9906, pp. 694–711. Springer, Cham (2016). https://doi.org/10.1007/978-3-319-46475-6_43
8. Karras, T., Laine, S., Aila, T.: A style-based generator architecture for generative adversarial networks. In: Proceedings of the IEEE Conference on Computer Vision and Pattern Recognition, pp. 4401–4410 (2019)
9. Nie, D., et al.: Medical image synthesis with deep convolutional adversarial networks. IEEE Trans. Biomed. Eng. **65**(12), 2720–2730 (2018)
10. Radford, A., Metz, L., Chintala, S.: Unsupervised representation learning with deep convolutional generative adversarial networks. arXiv preprint arXiv:1511.06434 (2015)
11. Reinhold, J.C., Dewey, B.E., Carass, A., Prince, J.L.: Evaluating the impact of intensity normalization on MR image synthesis. In: Medical Imaging 2019: Image Processing, vol. 10949, p. 109493H. International Society for Optics and Photonics (2019)
12. Ronneberger, O., Fischer, P., Brox, T.: U-Net: convolutional networks for biomedical image segmentation. In: Navab, N., Hornegger, J., Wells, W.M., Frangi, A.F. (eds.) MICCAI 2015. LNCS, vol. 9351, pp. 234–241. Springer, Cham (2015). https://doi.org/10.1007/978-3-319-24574-4_28
13. Shin, H.-C., et al.: Medical image synthesis for data augmentation and anonymization using generative adversarial networks. In: Gooya, A., Goksel, O., Oguz, I., Burgos, N. (eds.) SASHIMI 2018. LNCS, vol. 11037, pp. 1–11. Springer, Cham (2018). https://doi.org/10.1007/978-3-030-00536-8_1
14. Szegedy, C., Vanhoucke, V., Ioffe, S., Shlens, J., Wojna, Z.: Rethinking the inception architecture for computer vision. In: Proceedings of the IEEE Conference on Computer Vision and Pattern Recognition, pp. 2818–2826 (2016)

15. Tustison, N.J., et al.: N4ITK: improved N3 bias correction. IEEE Trans. Med. Imaging **29**(6), 1310–1320 (2010)
16. Wang, Z., Bovik, A.C., Sheikh, H.R., Simoncelli, E.P.: Image quality assessment: from error visibility to structural similarity. IEEE Trans. Image Process. **13**(4), 600–612 (2004)
17. Yi, X., Walia, E., Babyn, P.: Generative adversarial network in medical imaging: a review. Med. Image Anal. **58**, 101552 (2019)

Simulated Diffusion Weighted Images Based on Model-Predicted Tumor Growth

Pamela R. Jackson(✉), Andrea Hawkins-Daarud, and Kristin R. Swanson

Mayo Clinic, Phoenix, AZ 85054, USA
jackson.pamela@mayo.edu

Abstract. Non-invasive magnetic resonance imaging (MRI) is the primary imaging modality for visualizing brain tumor growth and treatment response. While standard MRIs are central to clinical decision making, advanced quantitative imaging sequences like diffusion weighted imaging (DWI) are increasingly relied on. Deciding the best way to interpret DWIs, particularly in the context of treatment, is still an area of intense research. With DWI being indicative of tissue structure, it is important to establish the link between DWI and brain tumor mathematical growth models, which could help researchers and clinicians better understand the tumor's microenvironmental landscape. Our goal was to demonstrate the potential for creating a DWI patient-specific untreated virtual imaging control (UVICs), which represents an individual tumor's untreated growth and could be compared with actual patient DWIs. We generated a DWI UVIC by combining a patient-specific mathematical model of tumor growth with a multi-compartmental MRI signal equation. GBM growth was mathematically modeled using the Proliferation-Invasion-Hypoxia-Necrosis-Angiogenesis-Edema (PIHNA-E) model, which simulated tumor as being comprised of multiple cellular phenotypes interacting with vasculature, angiogenic factors, and extracellular fluid. The model's output consisted of spatial volume fraction maps for each microenvironmental species. The volume fraction maps and corresponding T2 and apparent diffusion coefficient (ADC) values from literature were incorporated into a multi-compartmental signal equation to simulate DWI images. Simulated DWIs were created at multiple b-values and then used to calculate ADC maps. We found that the regional ADC values of simulated tumors were comparable to literature values.

Keywords: Biomathematical tumor growth model · Glioblastoma · Glioma · Simulated magnetic resonance imaging

1 Introduction

Glioblastoma (GBM) is the most common primary malignant brain tumor and is highly invasive and aggressive leaving patients with a poor median survival of 14.7-16.6 months [1]. Given the relative inaccessibility of brain tumor tissue, non-invasive magnetic resonance imaging (MRI) provides a necessary surrogate for assessing not only the extent of disease burden, but also treatment response [2]. Standard anatomical MRIs, including T2-weighted (T2W), T1-weighted (T1W), and T1W with gadolinium

N. Burgos et al. (Eds.): SASHIMI 2020, LNCS 12417, pp. 32–40, 2020.
https://doi.org/10.1007/978-3-030-59520-3_4

contrast agent (T1Gd), are regularly utilized in the clinic to visualize and monitor brain tumor growth. More recently, advanced quantitative imaging sequences, like diffusion weighted imaging (DWI), have become standard in brain tumor imaging protocols [3]. DWIs are sensitive to microscopic motion of water. Quantitative maps of *in vivo* diffusion, often referred to as the apparent diffusion coefficient (ADC), can be computed. For dense tissue structure like neoplasms, the general consensus is that higher ADC represents a lower cell density and lower ADC represents a high cell density, but the diffuse nature GBMs inherently means that this relationship may not be true everywhere within any given patient's tumor [4–7]. Further, how to interpret these images and maps in the context of treatment is still under review [8]. Utilizing biomathematical modeling could help decipher the complex microenvironmental evolution in the tumor. However, a clear link between tumor growth models and MRI signal intensities has not been established.

Simulating DWIs of the brain is an area of active research and provides some context for understanding the connection between structure and signal. Much research is driven by the complex nature of water diffusion and the microscopic scale of the biophysical processes that give rise to the DWI signal. Simulating DWIs is one way to represent subvoxel-resolution processes and to test methodologies for non-invasively quantifying diffusion *in vivo*. Work typically focuses on methods for generating phantom brain structure of varying complexity and then simulating the resulting signal evolution [9–13]. While there are numerous papers on simulating diffusion weighted images of the brain's current state, we found little evidence of simulated DWIs of the predicted brain tumor growth. In Prastawa et al., the averaged mean diffusivity (MD) and fractional anisotropy (FA) maps were modified based on the location of the simulated tumor's mass effect [14].

Our goal was to demonstrate the potential for creating a DWI patient-specific untreated virtual imaging control (UVICs), which represented an individual tumor's untreated growth and could be compared with actual patient DWIs. We generated a DWI UVIC by combining a patient-specific mathematical model of tumor growth with a multi-compartmental MRI signal equation, which was based on the Bloch equations. GBM growth was mathematically modeled using the Proliferation-Invasion-Hypoxia-Necrosis-Angiogenesis-Edema (PIHNA-E) model. We then compared the resulting ADC map from the example DWI UVIC to literature reported ADC values in different brain tumor regions. This work extended a previous simulation scheme for creating T2-weighted (T2W) MRIs [15] to create DWIs.

2 Diffusion-Weighted MRI

DWI is a MRI-based technique that probes the diffusive movement of water molecules in a given direction. Diffusion in this sense refers to self-diffusion or the translation of water due to Brownian motion [8]. In brain tissue, water diffuses in multiple environments including within the intracellular space (ICS), extracellular space (ECS), across cell membranes, and in the ventricles containing cerebrospinal fluid (CSF) [16]. The DWI signal intensity is related to diffusion in tissue according to

$$S_{DWI}(b) = S_{0(T2W)}e^{-bD} \tag{1}$$

where S_{DWI} is the diffusion weighted signal, $S_{0(T2W)}$ is the T2 weighted signal without diffusion weighting, b is the diffusion weighting value, and ADC is the apparent diffusion coefficient [17]. Larger b-values sensitize the method to diffusion and lead to lower signal intensities for faster diffusing regions. DWIs are typically acquired with at least two b-values for calculating ADC in a given direction using Eq. (1).

3 Methods

3.1 Creating Mock Tumors Through Mathematical Modeling

We utilized the previously published PIHNA-E model to create a mock simulated glioma with growth over time (Fig. 1). The full details of the PIHNA-E model can be found in [18]. Briefly, the model is initiated with a small number of cells allowed to grow on the BrainWeb MRI atlas [19, 20]. Tumor cells are modeled within an oxygen environment and can be normoxic, hypoxic, or necrotic. Normoxic and hypoxic cells lead to the aggregation of angiogenic factors, which also lead to the formation of leaky vasculature and thus vasogenic edema. Normal extracellular space was set at 20% and was allowed to vary, reaching up to 56%, as edema was generated [21–24].

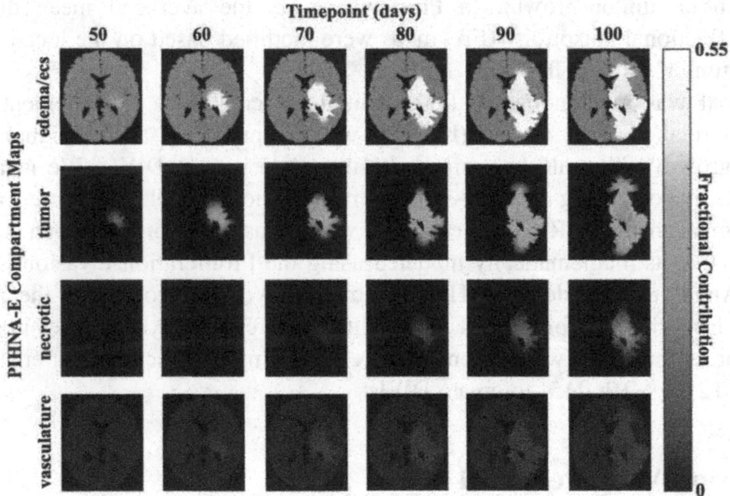

Fig. 1. The PIHNA-E model produces spatial maps for multiple tissue compartments. We utilized multiple growth timepoints to evaluate the simulated diffusion weighted MRIs. Adapted from [15].

3.2 Simulating DWIs Based on MRI Physics

Previously [15], we demonstrated creating T2W MRIs based on a closed form solution to the Bloch equations for the specific case of the spin echo sequence neglecting longitudinal relaxation effects

$$S_{T2W}(T_E) = S_0 e^{-T_E/T_2} \tag{2}$$

where S_{T2W} is the T2W signal, S_0 is the signal without any T2 weighting, TE is the echo time, and T2 is the transverse relaxation time [25]. Since the DWI is a modification to a standard MRI sequence, diffusion weighting was included by substituting Eq. (2) for $S_{0(T2W)}$ in Eq. (1) leading to

$$S_{DWI}(T_E, b) = S_0 e^{-T_E/T_2} e^{-bADC} \tag{3}$$

[17]. Equation (3) was then generalized for multiple compartments

$$S_{DWI}(T_E, b) = S_0 \sum_{n=1}^{N} f_n e^{-T_E/T_{2,n}} e^{-bADC_n} \tag{4}$$

where n is the compartment, N is the total number of compartments, f_n is the volume fraction for the nth compartment [26]. Each compartment was associated with a T2 value, ADC value, and a volume fraction.

Using the spatial maps from the PIHNA-E model and the simplified MRI signal equation, we simulated DWI MRIs assuming a spin-echo-based sequence (Eq. (4) and Fig. 2). We considered multiple compartments for the simulation: white matter cells, gray matter cells, tumor cells (hypoxic + normoxic cells), necrosis, cerebrospinal fluid (CSF), and extracellular space (white and gray extracellular fluid + edema), and vascular cells.

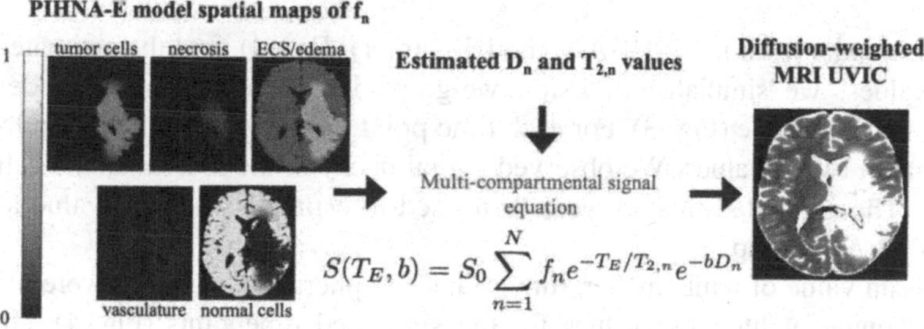

Fig. 2. Scheme for creating untreated virtual imaging control (UVIC) diffusion-weighted images. The spatial maps generated from the PIHNA-E biomathematical tumor growth model represent the volume fractions of each considered tissue compartment. The closed-form, multi-compartmental MRI signal equation utilizes the spatial PIHNA-E maps and associated T2s and ADC values from literature. Adapted from [15].

Definitions of white matter, gray matter, and CSF were based on the gray/white segmentation of the atlas. Except for the vascular compartment, all ADC values were chosen based on literature (Table 1). The value for vasculature was set to an extreme high value to represent vascular flow. T2 values were either estimated from our previous normal volunteer study [15] or were chosen based on literature as shown in Table 1. The values for ADC and T2 tumor cells were estimated to be the same as normal gray matter cells. We further simplified our calculations by assuming isotropic diffusion and measuring diffusion in one direction.

Table 1. Estimated compartmental values of ADC and T2 utilized to simulate DWIs.

Compartment	ADC ($\mu m^2/ms$)	T2 (ms)
White matter cells	465 [27]	66 [15]
Gray matter cells, tumor cells	465 [27]	80 [15]
White Matter ECS	1740 [28]	127 [15]
Gray Matter ECS	1740 [28]	166 [15]
CSF, Edema, necrosis	3000 [8]	2200 [15, 29]
Vasculature	1×10^5	68 [30]

3.3 Creating ADC Maps from Simulated DWIs

Simulated diffusion weighted images were created with $b = 0$ and $1000\ s/mm^2$, $TE = 110$ ms, and 185×147 matrix on 1 slice for each growth timepoint. The b and TE parameters were chosen to match those utilized in Oh et al., which reported ADC values for various regions of GBM [31]. ADC maps were calculated on a voxel-wise basis using Eq. (1). ROIs were placed in representative white matter, tumor, and peripheral edema regions to capture ADC values for comparison with Oh et al. [31].

4 Results

Using the outputs from the PIHNA-E model (Fig. 1), Eq. (4), and the estimated ADC and T2 values, we simulated diffusion weighted images representing model-based tumor growth over time (Fig. 3). For each time point shown in Fig. 3, we simulated the DWI signal at two b-values. We observed signal decay in the images as the echo time increased. The simulated images were then used to estimate the ADC value and displayed as an ADC map.

The mean value of white matter, tumor, and peripheral edema ROIs were primarily within the range of literature values for the simulated timepoints (Fig. 4). Oh et al. reported a range of tumor ADC values that were generally higher than peripheral edema. While simulated peripheral edema ADC values were largely in range, they were generally less than the simulated tumor ADC values.

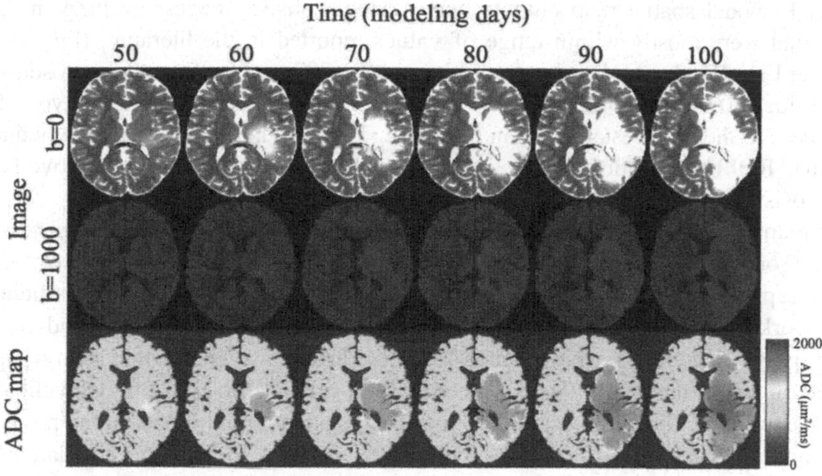

Fig. 3. Diffusion weighted UVICs with corresponding calculated ADC maps over time.

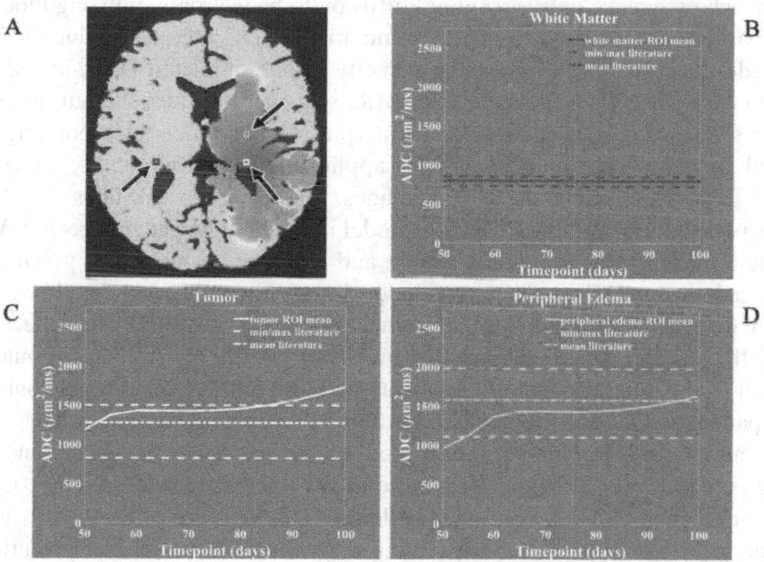

Fig. 4. A: ADC map with ROIs for evaluating white matter (red), tumor (white), and peripheral edema (cyan). Graphs showing ADC values over time for B: white matter, C: tumor, and D: peripheral edema. (Color figure online)

5 Discussion

In extending our previous method for generating simulated T2W MRIs to include the effects of diffusion weighting, we were able to simulate DWIs. Utilizing a multicompartmental MRI signal equation, ADC and T2 values derived from literature, and the

PIHNA-E model spatial map outputs, we created realistic images resulting in ADC values that were mostly within range of values reported in the literature (Fig. 4). As shown in Fig. 4B, the ADC value for white matter (797 $\mu m^2/ms$) was very close to the mean value (807 $\mu m^2/ms$) reported in Oh et al. [31]. The tumor ROI did have values that exceeded the range established in literature. These higher values were likely due to the tumor ROI being stationary across time and the region eventually being overtaken by necrosis (Fig. 1).

As many other efforts towards creating simulated DWIs recapitulate complex brain structure, our method represents a minimalistic perspective where the subvoxel regime was conceptualized as simple volumetric compartments rather than complex structures. Other works simulating DWIs mainly focused on simulating images based on the anisotropy of the brain's white matter [9–12]. Further, we found that many papers focused on the normal brain [9–11], ischemia [12], cell swelling [13], or demyelination [32]. Prastwa et al. did demonstrate MD and FA maps modified based on the predicted brain tumor's mass effect, however, their work was not focused on simulating the underlying images that are used to calculate those maps [14]. To our knowledge, there is not a previous publication simulating a predicted brain tumor's DWI.

Limitations of this work include creating images representative of the pulsed gradient spin echo sequence only, assuming diffusion to be isotropic, utilizing mock data, and comparing our simulated results to summarized ADC literature values. We have planned additional work to increase the veracity of our simulated DWI, including the ability to create images for other types of MRI sequences besides the pulsed gradient spin echo sequence (e.g. oscillating gradient spin echo). Models incorporating anisotropy will increase the simulated DWI's applicability and our ability to compare simulated images more directly to real images. Our current work does not depict a particular patient, however, the PIHNA-E model is meant to be patient-specific. We can instantiate a model for an individual patient and compare values to that given patient. The Cancer Imaging Archive (TCIA) includes diffusion imaging and could be a source of patient images. Additionally, there is an effort at our institution to collect multispectral MRIs as part of an image-guided biopsy effort, and those images could be of use as well [33]. Utilizing a cohort of patients will also allow us to compare our results on a per patient as well as a group basis. Further, one could certainly take a data-driven approach and train a machine learning model with the PIHNA-E maps and corresponding patient DWIs, allowing for a more direct link between the PIHNA-E model and real-world images. However, machine learning models are often viewed as black boxes due to a lack of explainability. Using a mechanistic model can potentially be more flexible, enabling simulations of a specific set of parameters that match those of a real or desired image.

6 Conclusion

To demonstrate the concept of DWI patient-specific UVICs, we simulated DWI images based on the PIHNA-E biomathematical brain tumor model. Utilizing characteristic values for various tissue compartments represented in the tumor model and a multi-compartmental MRI signal equation, realistic images were created. Calculating ADC maps based on DWIs at two different b values, we were able to produce ADC values in regions of the brain and tumor that were comparable to literature.

Acknowledgements. The authors would like to thank Drs. Savannah C. Partridge and Paul E. Kinahan for many helpful discussions. Further, we acknowledge the following funding sources that made our work possible: John M. Nasseff, Sr. Career Development Award in Neurologic Surgery Research, Moffitt PSOC U54CA193489, Diversity Supplement 3U54CA193489-04S3, Mayo U01 U01CA220378, and MIT PSOC U54CA210180.

References

1. Stupp, R., et al.: European organisation for research and treatment of cancer brain tumor and radiotherapy groups, national cancer institute of canada clinical trials group: radiotherapy plus concomitant and adjuvant temozolomide for glioblastoma. N. Engl. J. Med. **352**, 987–996 (2005)
2. Clarke, J.L., Chang, S.M.: Neuroimaging: diagnosis and response assessment in glioblastoma. Cancer J. **18**, 26–31 (2012)
3. Schmainda, K.M.: Diffusion-weighted MRI as a biomarker for treatment response in glioma (2012). https://doi.org/10.2217/cns.12.25
4. Sugahara, T., et al.: Usefulness of diffusion-weighted MRI with echo-planar technique in the evaluation of cellularity in gliomas. J. Magn. Reson. Imaging **9**, 53–60 (1999)
5. Ellingson, B.M., et al.: Validation of functional diffusion maps (fDMs) as a biomarker for human glioma cellularity. J. Magn. Reson. Imaging **31**, 538–548 (2010)
6. Gupta, R.K., et al.: Relationships between choline magnetic resonance spectroscopy, apparent diffusion coefficient and quantitative histopathology in human glioma. J. Neurooncol. **50**, 215–226 (2000). https://doi.org/10.1023/A:1006431120031
7. Eidel, O., et al.: Automatic analysis of cellularity in glioblastoma and correlation with ADC using trajectory analysis and automatic nuclei counting. PLoS ONE **11**, e0160250 (2016)
8. Le Bihan, D., Iima, M.: Diffusion magnetic resonance imaging: what water tells us about biological tissues. PLoS Biol. **13**, e1002203 (2015)
9. Neher, P.F., Laun, F.B., Stieltjes, B., Maier-Hein, K.H.: Fiberfox: facilitating the creation of realistic white matter software phantoms. Magn. Reson. Med. **72**, 1460–1470 (2014)
10. Perrone, D., et al.: D-BRAIN: anatomically accurate simulated diffusion MRI brain data. PLoS ONE **11**, e0149778 (2016)
11. Close, T.G., Tournier, J.-D., Calamante, F., Johnston, L.A., Mareels, I., Connelly, A.: A software tool to generate simulated white matter structures for the assessment of fibre-tracking algorithms. Neuroimage **47**, 1288–1300 (2009)
12. Harkins, K.D., Galons, J.-P., Secomb, T.W., Trouard, T.P.: Assessment of the effects of cellular tissue properties on ADC measurements by numerical simulation of water diffusion (2009). https://doi.org/10.1002/mrm.22155
13. Yeh, C.-H., Schmitt, B., Le Bihan, D., Li-Schlittgen, J.-R., Lin, C.-P., Poupon, C.: Diffusion microscopist simulator: a general Monte Carlo simulation system for diffusion magnetic resonance imaging. PLoS ONE **8**, e76626 (2013)
14. Prastawa, M., Bullitt, E., Gerig, G.: Simulation of brain tumors in MR images for evaluation of segmentation efficacy. Med. Image Anal. **13**, 297–311 (2009)
15. Jackson, P.R., Hawkins-Daarud, A., Partridge, S.C., Kinahan, P.E., Swanson, K.R.: Simulating magnetic resonance images based on a model of tumor growth incorporating

microenvironment. In: Medical Imaging 2018: Image Perception, Observer Performance, and Technology Assessment, p. 105771D (2018)
16. Le Bihan, D., Johansen-Berg, H.: Diffusion MRI at 25: exploring brain tissue structure and function. Neuroimage **61**, 324–341 (2012)
17. Stejskal, E.O., Tanner, J.E.: Spin diffusion measurements: spin echoes in the presence of a time-dependent field gradient (1965). https://doi.org/10.1063/1.1695690
18. Hawkins-Daarud, A., Rockne, R.C., Anderson, A.R.A., Swanson, K.R.: Modeling tumor-associated edema in gliomas during anti-angiogenic therapy and its impact on imageable tumor. Front. Oncol. **3**, 66 (2013)
19. Collins, D.L., et al.: Design and construction of a realistic digital brain phantom. IEEE Trans. Med. Imaging **17**, 463–468 (1998)
20. BrainWeb: Simulated brain database. http://www.bic.mni.mcgill.ca/brainweb/. Accessed 28 June 2020
21. Nicholson, C., Syková, E.: Extracellular space structure revealed by diffusion analysis. Trends Neurosci. **21**, 207–215 (1998)
22. Xiao, F., Nicholson, C., Hrabe, J., Hrabetová, S.: Diffusion of flexible random-coil dextran polymers measured in anisotropic brain extracellular space by integrative optical imaging. Biophys. J. **95**, 1382–1392 (2008)
23. Vargová, L., Homola, A., Zámecník, J., Tichý, M., Benes, V., Syková, E.: Diffusion parameters of the extracellular space in human gliomas. Glia **42**, 77–88 (2003)
24. Bruehlmeier, M., et al.: Measurement of the extracellular space in brain tumors using 76Br-bromide and PET. J. Nucl. Med. **44**, 1210–1218 (2003)
25. Hahn, E.L.: Spin echoes (1950). https://doi.org/10.1063/1.3066708
26. Smouha, E., Neeman, M.: Compartmentation of intracellular water in multicellular tumor spheroids: diffusion and relaxation NMR. Magn. Reson. Med. **46**, 68–77 (2001)
27. Clark, C.A., Hedehus, M., Moseley, M.E.: In vivo mapping of the fast and slow diffusion tensors in human brain. Magn. Reson. Med. **47**, 623–628 (2002)
28. Maier, S.E., Mulkern, R.V.: Biexponential analysis of diffusion-related signal decay in normal human cortical and deep gray matter. Magn. Reson. Imaging **26**, 897–904 (2008)
29. Brown, R.W., Mark Haacke, E.-C., Norman Cheng, Y., Thompson, M.R., Venkatesan, R.: Magnetic Resonance Imaging: Physical Principles and Sequence Design. Wiley, Hoboken (2014)
30. Krishnamurthy, L.C., Liu, P., Ge, Y., Lu, H.: Vessel-specific quantification of blood oxygenation with T2-relaxation-under-phase-contrast MRI. Magn. Reson. Med. **71**, 978–989 (2014)
31. Oh, J., et al.: Quantitative apparent diffusion coefficients and T2 relaxation times in characterizing contrast enhancing brain tumors and regions of peritumoral edema. J. Magn. Reson. Imaging **21**, 701–708 (2005)
32. Salan, T., Jacobs, E.L., Reddick, W.E.: A 3D model-based simulation of demyelination to understand its effects on diffusion tensor imaging. In: Conference Proceedings of IEEE Engineering Medical Biology Society, pp. 3525–3528 (2017)
33. Hu, L.S., et al.: Accurate patient-specific machine learning models of glioblastoma invasion using transfer learning (2019). https://doi.org/10.3174/ajnr.a5981

Blind MRI Brain Lesion Inpainting Using Deep Learning

José V. Manjón[1(✉)], José E. Romero[1], Roberto Vivo-Hernando[2], Gregorio Rubio[3], Fernando Aparici[4], Maria de la Iglesia-Vaya[5], Thomas Tourdias[6], and Pierrick Coupé[7]

[1] Instituto de Aplicaciones de las Tecnologías de la Información y de las Comunicaciones Avanzadas (ITACA), Universitat Politècnica de València, Camino de Vera s/n, 46022 Valencia, Spain
jmanjon@fis.upv.es

[2] Instituto de Automática e Informática Industrial, Universitat Politècnica de València, Camino de Vera s/n, 46022 Valencia, Spain

[3] Departamento de Matemática Aplicada, Universitat Politècnica de València, Camino de Vera s/n, 46022 Valencia, Spain

[4] Área de Imagen Médica, Hospital Universitario y Politécnico La Fe, Valencia, Spain

[5] Brain Connectivity Laboratory, Joint Unit FISABIO and Prince Felipe Research Centre (CIPF), Valencia, Spain

[6] CHU de Bordeaux, service de neuroimagerie diagnostique et thérapeutique, 33076 Bordeaux, France

[7] CNRS, LaBRI, UMR 5800, PICTURA, 33400 Talence, France

Abstract. In brain image analysis many of the current pipelines are not robust to the presence of lesions which degrades their accuracy and robustness. For example, performance of classic medical image processing operations such as non-linear registration or segmentation rapidly decreases when dealing with lesions. To minimize their impact, some authors have proposed to inpaint these lesions so classic pipelines can be used. However, this requires to manually delineate the regions of interest which is time consuming. In this paper, we propose a deep network that is able to blindly inpaint lesions in brain images automatically allowing current pipelines to robustly operate under pathological conditions. We demonstrate the improved robustness/accuracy in the brain segmentation problem using the SPM12 pipeline with our automatically inpainted images.

Keywords: Lesion inpainting · MRI · Deep learning · Robust segmentation

1 Introduction

Most of state-of-the-art brain image analysis packages have been designed to work with anatomically consistent images using some assumptions such as the existence of a fixed number of expected tissues for example. Therefore, when the images have pathological alterations such as lesions, these assumptions are violated leading to biased/incorrect results. Although the development of more complex pipelines covering a wider set of

N. Burgos et al. (Eds.): SASHIMI 2020, LNCS 12417, pp. 41–49, 2020.
https://doi.org/10.1007/978-3-030-59520-3_5

conditions could manage this problem, we are also interested by using well-established standard packages such as Freesurfer or SPM (for example for longitudinal analysis).

One possible solution to this problem has been already suggested by several authors in the past, image inpainting. Image inpainting [1] refers to the process where specific regions of an image or volume are masked-out and filled with plausible information from the image itself or other "similar" images. In medical imaging, this technique has been used, for example, to fill multiple sclerosis (MS) lesions with white matter consistent intensities resulting in realistic healthy-looking versions of the images [2–4]. Guizard et al. [3] filled the lesion regions with locally consistent intensities using a patch-based multi-resolution onion peel strategy. More recently, Prados et al. [4] proposed a "modality-agnostic" patch-based method for lesion filling also based on the non-local means strategy. Using deep learning, Armanious et al. [5] recently proposed a 2D network to inpaint medical image modalities using an adversarial approach.

All these inpainting methods require the definition of the regions to inpaint. This definition can be performed manually (which is a time consuming, tedious and error prone procedure) or automatically using a lesion segmentation algorithm. Unfortunately, segmentation-based approaches highly depend on the quality of the segmentation (especially the number of false negatives).

An alternative where there is no need to explicitly define the regions to inpaint is called blind inpainting [6]. Blind inpainting behaves as a restoration process where the input image is considered as "corrupted" and the method tries to directly predict the "clean" image by learning the relationship between them. For example, Liu et al. [6] used a Residual Deep network to predict the clean image without any specific definition on the regions to fill.

In this paper, we propose, as far as we know, the first 3D blind inpainting method in medical imaging. We propose a deep network which is able to predict a "clean" image from an image containing abnormal areas (lesions). To construct our network, we split the process into two steps. We first learn how to inpaint anatomically normal images (with random regions masked out) and we applied it to reconstruct MS images masking out the lesions to obtain their healthy versions. In the second step, we use these healthy-looking images as output of a network feed with their corresponding unhealthy version. We show the benefits of our inpainting method on segmentation problem but its application is not limited to this problem as it can help in other problems such as longitudinal non-linear registration for example.

2 Materials and Methods

2.1 Datasets

We used three different datasets, one containing cases with no anatomical abnormalities (lesions) and two with multiple sclerosis (MS) lesions.

Normal Dataset Without Lesions: The open access IXI dataset (https://brain-development.org/ixi-dataset/) was used. This dataset contains images of 580 healthy subjects from several hospitals in London (UK). Both 1.5T and 3T images were included in our training dataset. 3T images were acquired on a Philips Intera 3T scanner (TR = 9.6 ms, TE = 4.6 ms, flip angle = 8°, slice thickness = 1.2 mm, volume size = 256 × 256 × 150, voxel dimensions = 0.94 × 0.94 × 1.2 mm^3). 1.5 T images were acquired on a Philips Gyroscan 1.5T scanner (TR = 9.8 ms, TE = 4. 6 ms, flip angle × = × 8°, slice thickness = 1.2 mm, volume size = 256 × 256 × 150, voxel dimensions = 0.94 × 0.94 × 1.2 mm^3). Images from subjects presenting visual lesions or severe image artifacts were excluded from selection, being the final number of subjects used of 298.

MS Dataset with Manual Segmentation of MS Lesions: Our MS dataset is composed of 3T 3D-T1w MPRAGE images from 43 patients previously used in the development of lesionBrain pipeline [7]. These images were preprocessed to align them into the MNI space and to normalize their intensities. Manual segmentations of white matter lesions were obtained as described in [7].

MS Dataset with Automatic Segmentation of MS Lesions: This dataset consist of 200 cases automatically analyzed using lesionBrain segmentation pipeline within the volBrain online platform. These T1 + lesion masks were visually checked to assess their quality.

2.2 Pipeline Description

Preprocessing: First, the images are preprocessed to normalize their intensities and to register them into the MNI space (1 mm^3 voxel resolution). A denoising step based on the adaptive nonlocal means filter is first applied [8]. Next, the images are registered into the MNI space using an affine transform [9]. Finally, inhomogeneity correction is performed [10]. Intensity standardization was performed dividing the images by their mean.

Inpainting Network Topology: The proposed network is a 3D UNet-like network [11] (see Fig. 1), which learns a non-linear mapping between the source image (image to inpaint) and target image (inpainted image). To work with the full 3D volume, due to memory constraints, we decomposed the input volume into 8 sub-volumes using a 3D spatial-to-depth decomposition by decimating the original volume with one voxel shift in each 3D dimension. These 8 sub-volumes were structured as an input tensor of 91 × 109 × 91 × 8 containing all the voxels of the original input volume.

In our 3D UNET, the encoder path consists of a 3D convolution layer (kernel size of 3 × 3 × 3 voxels) for each resolution level with ReLU activation and batch normalization layers followed by a max pooling layer and upsampling layer is used in the decoder path. The first resolution level used 40 filters with an increment of factor 2 in the subsequent lower resolution levels in order to balance the loss of spatial resolution. Similarly, the number of filters is decreased by 2 in the decoder path at subsequent resolution levels. We added a final reconstruction block consisting in a tri-linear interpolation layer followed by a 3D convolution layer (with 8 filters) plus a.

ReLU and Batch normalization layers. Finally, a last 3D convolution layer with a single filter of size 1 × 1 × 1 was used to produce the full resolution 3D output (see Fig. 1 for more details).

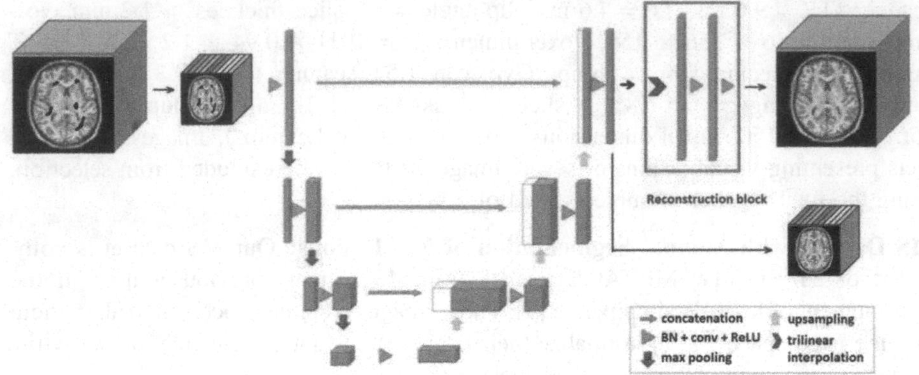

Fig. 1. Scheme of the proposed 3D network for MRI inpainting.

To train the network, we used the mean squared error loss and a deep supervision approach [12]. The first output is the low-resolution depth-wise output and the second the full resolution volume after the reconstruction block (see Fig. 1).

Non-blind vs Blind Inpainting: In classic inpainting procedures one or several regions are marked out in the image as areas to be filled with "realistic" data from other images or the image itself. Therefore, this requires the explicit manual definition of these areas which is time consuming and error prone. In this paper, we propose a deep convolution neural network for blind WM lesion inpainting that does not require the manual definition of lesions. To obtain this network, we split the process in two steps: i) generation of synthetic training data using a non-blind inpainting network and ii) the development of a blind inpainting network.

1. **Development of a non-blind inpainting network.** In order to generate training dataset for the blind inpainting network described in the following, we first train a non-blind inpainting network able to reconstruction healthy looking image from masked-out images. To train this network, we used the architecture previously defined. As training data, we used as input randomly selected normal anatomy cases (N = 298) masked out with randomly selected lesion masks (N = 500) automatically generated using lesionBrain (which includes large, medium and small lesion load examples). Note that these lesion masks were different from the masks of the MS dataset with automatic lesion segmentation. As output, we used the original unmasked images. We masked out the regions to inpaint by setting those pixels to −1 to clearly differentiate them from the pixels that have to remain unchanged that are positive defined.
2. **Development of a blind inpainting network**: To train this network, we used the same architecture but with different training data. As input data, we used automatically segmented MS images without any masking (N = 200). Since no lesion-free version of our MS cases is available for the output, we used as a surrogate their inpainted versions obtained by applying the previously trained non-blind inpainting network masking out the lesions using the available automatic segmentation.

To teach the network also when not to inpaint (healthy cases) we added as input/output images from IXI dataset. In this case the network just behaves as an autoencoder. This lets the network to be blindly applied seamlessly to either healthy or lesioned cases.

3 Results

All the experiments were performed using Tensorflow 1.15 and Keras 2.2.4 in a Titan X GPU with 12 GB of memory. To train the network we used an Adam optimizer with default parameters. A data generator function was used to feed the network during 200 epochs (50 cases per epoch). To measure the performance of the proposed method, we used two different metrics providing information from different perspectives, the peak-signal-to-noise ratio (PSNR) and the correlation coefficient (all computed over the whole volume).

Reconstruction Performance
First, we trained the non-blind inpainting network. We randomly selected from the normal dataset 258 cases for training, 20 for validation and 20 for test. We used data augmentation within the generator consisting of left-right flipping the input and output randomly. The test PSNR was 48.13 vs. 34.21 of the input images and the correlation was almost perfect 0.9997 compared to the original input image correlation (0.9835). Moreover, a careful visual inspection corroborated the successful recovery of WM areas. Therefore, we considered our first non-blind inpainting network able to reconstruct healthy-looking images from masked-out images. This non-blind network is then used to generate synthetic training data for the blind inpainting network. Consequently, we applied this network to the automatically segmented MS dataset to generate the corresponding lesion-free images. Again, a visual inspection verified the successful generation of the images.

Second, we trained the blind inpainting network using as input the original MS data and as output the inpainted images generated by the first network. This time, we randomly selected 180 cases for training and 20 for validation. The test images were the original 43 manually segmented cases. We also added to the training dataset the 268 normal cases (218 for training, 20 for validation and 20 for test) so the network can also learn when to inpaint and when not do it. Again, we used a data augmentation policy within the generator consisting of left-right flipping the input and output randomly. The test PSNR for the blind network was 49.57 for MS cases and 44.97 for the normal images. The correlation of both test MS cases (0.9998) and normal cases (0.9999) was again very high.

From the efficiency point of view, the proposed method has a computational time of 0.5 s when running in GPU and around 8 s in CPU.

Impact on Segmentation

To evaluate the impact of the inpainting process in the segmentation of brain tissues on normal subjects and MS patients, we used the well-known SPM12 software [13]. SPM12 uses a generative model to segment the brain into different tissues such as grey matter (GM), white matter (WM) and cerebrum-spinal fluid (CSF) among others. Although SPM12 uses a powerful Bayesian approach including spatial tissue priors to help in the classification process, it normally fails when WM lesions are present, incorrectly labeling them as GM. Eventually, these lesions can be masked out to avoid problems, but this requires their manual or automatic delineation. Alternatively, we propose to segment the blindly inpainted images.

We segmented the non-blind inpainted (first network) MS test images with SPM12 and we considered them as the ground truth segmentations after checking that they did not include any WM lesion misclassification. Then SPM12 was also applied to the original and blind-inpainted images and their segmentations were compared to the ground truth using the DICE metric. Results are summarized in Table 1 and an example result is shown at Fig. 2.

Table 1. Mean Dice comparison on test MS data using the proposed blind-inpainting network. *Statistically significant differences.

Tissue	MS (N = 43)	
	Original	Inpainted
GM	0.9729 ± 0.0158	**0.9845 ± 0.0115***
WM	0.9707 ± 0.0190	**0.9823 ± 0.0141***
Average	0.9718 ± 0.0172	**0.9834 ± 0.0128***

As can be noted, the inpainted MS images provided significantly better DICE coefficients in average and for both GM and WM tissues (Wilcoxon signed-rank test).

Abnormal WM Segmentation

Although the inpainted images can be used to robustly classify GM and WM tissues, we might want to label the WM lesions as abnormal WM since this can be an interesting information for some neurological diseases. A simple way to obtain this information is to compute the absolute difference between the original and inpainted images and then select those voxels (within the WM area) with values higher than a fixed thresholdλ. An illustrative example of this approach ($\lambda = 0.5$) is shown in Fig. 3.

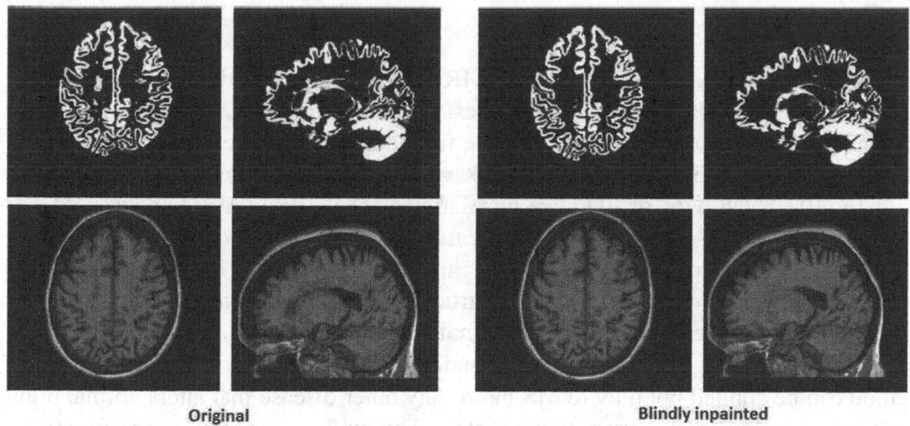

Fig. 2. Left: Original MS image and its GM segmentation. Right: Blindly-inpainted image and its GM segmentation (note that most of WM lesions no longer appear as GM).

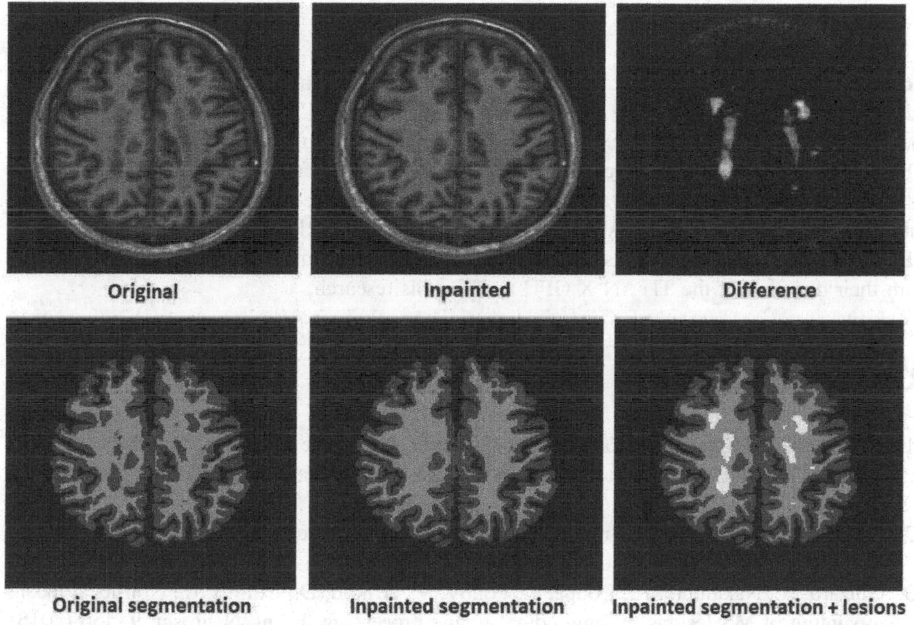

Fig. 3. First row: example of abnormal WM segmentation obtained from the image difference between original and inpainted images. Second row: segmentation results.

4 Conclusion

In this paper, we have presented a novel MRI 3D inpainting method which is able to blindly remove WM lesions from T1 images. As far as we know, this is the first blind inpainting method proposed in MRI and the first fully 3D approach (previous methods work only in 2D or 3D patches). To work with a single volume covering the entire brain (no 3D patch-wise or 2D slice-wise strategies) we used a 3D spatial-to-depth decomposition which allows to use a larger number of filters without losing any image resolution. To construct our network, we first learned how to inpaint anatomically normal images and we applied it to reconstruct MS images masking out the lesions to obtain their healthy versions. Finally, our final inpainting network was obtained simply mapping the MS images in their healthy versions. It is worth to note that our proposed method can be applied not only to MS but to any other disease that alters normal brain anatomy (vascular lesions, brain tumors, etc.). This will be explored in the future.

We have shown the benefits of our inpainting method on the segmentation problem using as example SPM12 software where robust tissue segmentation can be obtained (and outlier detection, i.e. lesions). However, its usefulness is not limited to this specific problem as it can help in other problems such as longitudinal non-linear registration for example.

Acknowledgement. This research was supported by the Spanish DPI2017-87743-R grant from the Ministerio de Economia, Industria y Competitividad of Spain. This work also benefited from the support of the project DeepvolBrain of the French National Research Agency (ANR-18-CE45-0013). This study was achieved within the context of the Laboratory of Excellence TRAIL ANR-10-LABX-57 for the BigDataBrain project. Moreover, we thank the Investments for the future Program IdEx Bordeaux (ANR-10- IDEX- 03- 02, HL-MRI Project), Cluster of excellence CPU and the CNRS. The authors gratefully acknowledge the support of NVIDIA Corporation with their donation of the TITAN X GPU used in this research.

References

1. Bertalmio, M., Sapiro, G., Caselles, V., Ballester, C.: Image inpainting. In: Proceedings of the 27th Annual Conference on Computer Graphics and Interactive Techniques, pp. 417–424 (2000)
2. Sdika, M., Pelletier, D.: Non rigid registration of multiple sclerosis brain images using lesion inpainting for morphometry or lesion mapping. Hum. Brain Mapp. **30**, 1060–1067 (2009)
3. Guizard, N., Nakamura, K., Coupé, P., Fonov, V., Arnold, D., Collins, L.: Non-local means inpainting of MS lesions in longitudinal image processing. Front. Neurosci. **9**, 456 (2015)
4. Prados, F., Cardoso, M.J., Kanber, B., et al.: A multi-time-point modality-agnostic patch-based method for lesion filling in multiple sclerosis. Neuroimage **139**, 376–384 (2016)
5. Armanious, K., Mecky, Y., Gatidis, S., Yang, B.: Adversarial inpainting of medical image modalisties. In: ICASSP2019 (2019)
6. Liu, Y., Pan, J., Su, Z.: Deep blind image inpainting. In: Cui, Z., Pan, J., Zhang, S., Xiao, L., Yang, J. (eds.) IScIDE 2019. LNCS, vol. 11935, pp. 128–141. Springer, Cham (2019). https://doi.org/10.1007/978-3-030-36189-1_11

7. Coupé, P., Tourdias, T., Linck, P., Romero, J.E., Manjón, J.V.: LesionBrain: an online tool for white matter lesion segmentation. In: Bai, W., Sanroma, G., Wu, G., Munsell, B.C., Zhan, Y., Coupé, P. (eds.) Patch-MI 2018. LNCS, vol. 11075, pp. 95–103. Springer, Cham (2018). https://doi.org/10.1007/978-3-030-00500-9_11
8. Manjon, J.V., et al.: Adaptive non-local means denoising of MR images with spatially varying noise levels. J. Magn. Reson. Imaging **31**(1), 192–203 (2010)
9. Avants, B.B., et al.: A reproducible evaluation of ANTs similarity metric performance in brain image registration. Neuroimage **54**(3), 2033–2044 (2011)
10. Tustison, N.J., et al.: N4ITK: improved N3 bias correction. IEEE Trans. Med. Imaging **29** (6), 1310–1320 (2010)
11. Ronneberger, O., Fischer, P., Brox, T.: U-Net: convolutional networks for biomedical image segmentation. In: Navab, N., Hornegger, J., Wells, William M., Frangi, Alejandro F. (eds.) MICCAI 2015. LNCS, vol. 9351, pp. 234–241. Springer, Cham (2015). https://doi.org/10.1007/978-3-319-24574-4_28
12. Dou, Q., Yu, L., Chen, H., Jin, Y., Yang, X., Qin, J., Heng, P.A.: 3D deeply supervised network for automated segmentation of volumetric medical images. Med. Image Anal. **41**, 40–54 (2017)
13. Ashburner, J.: SPM: a history. Neuroimage **62**(2), 791–800 (2012)

High-Quality Interpolation of Breast DCE-MRI Using Learned Transformations

Hongyu Wang[1,2(✉)], Jun Feng[3], Xiaoying Pan[1,2], Di Yang[4], and Baoying Chen[5(✉)]

[1] School of Computer Science and Technology, Xi'an University of Posts and Telecommunications, Xi'an 710121, Shaanxi, China
hywang@xupt.edu.cn

[2] Shaanxi Key Laboratory of Network Data Analysis and Intelligent Processing, Xi'an University of Posts and Telecommunications, Xi'an 710121, Shaanxi, China

[3] Department of Information Science and Technology, Northwest University, Xi'an 7101127, Shaanxi, China

[4] Functional and Molecular Imaging Key Lab of Shaanxi Province, Department of Radiology, Tangdu Hospital, Fourth Military Medical University, Xi'an 710038, Shaanxi, China

[5] Imaging Diagnosis and Treatment Center, Xi'an International Medical Center Hospital, Xi'an 710110, Shaanxi, China
chenby128@163.com

Abstract. Dynamic Contrast Enhancement Magnetic Resonance Imaging (DCE-MRI) is gaining popularity for computer aided diagnosis (CAD) of breast cancer. However, the performance of these CAD systems is severely affected when the number of DCE-MRI series is inadequate, inconsistent or limited. This work presents a High-Quality DCE-MRI Interpolation method based on Deep Neural Network (HQI-DNN) using an end-to-end trainable Convolutional Neural Network (CNN). It gives a good solution to the problem of inconsistent and inadequate quantity of DCE-MRI series for breast cancer analysis. Starting from a nested CNN for feature learning, the dynamic contrast enhanced features of breast lesions are learned by bidirectional contrast transformations between DCE-MRI series. Each transformation contains the spatial deformation field and the intensity change, enabling a variable-length multiple series interpolation of DCE-MRI. We justified the proposed method through extensive experiments on our dataset. It produced a more efficient result of breast DCE-MRI interpolation than other methods.

Keywords: DCE-MRI · Interpolation · CNN · Breast cancer

1 Introduction

Breast cancer, one of the leading cancers worldwide, is the main cause of death among woman who have been diagnosed with cancers. Effective breast cancer

N. Burgos et al. (Eds.): SASHIMI 2020, LNCS 12417, pp. 50–59, 2020.
https://doi.org/10.1007/978-3-030-59520-3_6

diagnosis in its early stage helps to save lives [1]. Dynamic Contrast Enhanced Magnetic Resonance Imaging (DCE-MRI), which produces high sensitivity and high resolution in dense breast tissues, is a valuable diagnostic tool for breast cancer [2]. Depending on the required precision and the allowed imaging time, DCE-MRI is acquired in a number of series. A typical breast DCE-MRI study contains dynamic contrast-enhanced series before and after the injection of a contrast agent at different time points.

However, it is a complex and time-consuming task for manual diagnosis due to the high dimension of 4D DCE-MRI [3]. Recent studies have shown that computer-aided diagnosis (CAD) methods are able to provide faster and reproducible results. Some CAD methods based on Convolutional Neural Networks (CNNs) achieve processing results of breast cancer diagnosis with DCE-MRI [4]. However, the number of DCE-MRI series used in these CAD methods is different, such as six DCE-MRI series used in [5], five DCE-MRI series used in [6] and only one DCE-MRI series used in [7]. The difference in DCE-MRI series is common to the clinic. Limited DCE-MRI series carry limited information. Most researches only use a few DCE-MRI series due to the difficulty of data collection, transmission and processing. In this sense, the inconsistency and inadequacy of DCE-MRI series severely constrains the performance of CAD systems.

In breast cancer analysis, deep learning-based CAD systems tend to produce state-of-the-art results when sufficient training data is offered. Data augmentation is commonly performed to boost the ability of deep learning network to learn features of images by increasing the number of training examples [8]. The current mainstream approach for data augmentation is divided into two main categories. (1) Hand-tuned approach. Some data augmentation functions (e.g., rotation, flip, zoom, shift, scale, contrast enhancement) are utilized to generate more additional learning samples. However, these functions have limited ability to emulate real variations of DCE-MRI, and they are very sensitive to the choice of parameters. (2) Generative Adversarial Networks (GAN) [9]. It can generate many simulated images by the joint training of a generator and a discriminator. Although there is ongoing research trying to improve lesion diagnosis accuracy with GAN, they do not explicitly account for variation in the kinetics behavior of the lesions among breast DCE-MRI series.

To alleviate the above issues, recent works have proposed learning data augmentation from raw data [10]. For example, Zhao et al. [11] proposed to model the spatial and appearance transformations between the labeled and unlabeled MRI brain scans, and produce a wide variety of realistic new images. They showed that training a supervised segmentation method with these new examples outperforms existing segmentation methods. But it can only generate single-frame image, not arbitrary multi-series DCE-MRI. Other recent works computing synthesized frames of a video focus on handling large object motion or occlusion [12,13]. These methods usually estimate dense motion correspondence between consecutive video frames via stereo matching or optical flow prediction [14]. However, the visual differences of tissue are very subtle compared with natural images [11].

Inspired by the learning-based data augmentation and video interpolation, we present a variable-length multiple series interpolation method for breast DCE-MRI. Our main idea is to warp the input two DCE-MRI series to the specific time step and then adaptively fuse the two warped series to generate the intermediate multiple DCE-MRI series. The kinetic and spatial transformation are modeled in an end-to-end trainable CNN network. Our contributions are: (1) To the best of our knowledge, this is the first deep learning framework that learns the kinetic and spatial transformations from DCE-MRI series for breast DCE-MRI interpolation. (2) Our investigation on how hierarchical nested-Unet estimates the blood flows reveals that spatial-temporal features play a significant role in DCE-MRI series interpolation. It can estimate breast DCE-MRI at arbitrary time step between DCE-MRI series, while existing works do not provide such results. (3) The proposed HQI-DNN framework achieves a better interpolation performance than other methods.

2 Materials and Methods

As shown in Fig. 1, the proposed HQI-DNN contains the following steps: (A) Extract the region of interests (ROIs) from DCE-MRI. (B) Given two input series of a lesion ROI, the contrast transformation model is learned to capture the kinetic behavior in the lesions. Specifically, the forward and backward contrast transformations are learned by a nested-Unet. Then, an appearance estimation module is proposed to generate the intermediate DCE-MRI series.

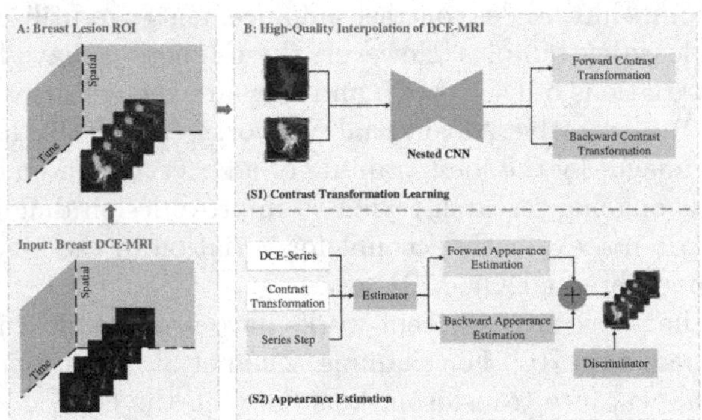

Fig. 1. An overview of the proposed framework, High-Quality Interpolation of breast DCE-MRI based on Deep Neural Network (HQI-DNN).

2.1 Dataset

The DCE-MRI dataset was acquired using a GEDiscoveryMR750 3.0T MRI system, which includes DCE-MRI of 59 female patients at a high risk of breast cancer. The contrast agent was injected through the dorsal vein of the hand from a high-pressure syringe (does of 0.2 mmol/kg, flow rate 2.0 ml/s). The interval between the consecutive scans was 60 s. The repetition time was 3.9 ms and the echo time was the minimum. Two experienced breast radiologists made pixel-wise labels on 1223 axial slices, and each slice contains one pre-contrast series and five post-contrast series. Firstly, the ROIs are obtained by a manual segmentation. All lesions are resized to $r \times r(r = 224)$ pixels with the bi-linear interpolation. The dataset is represented as $\mathbb{D} = \{(P_i\}_{i=1}^{|M|}$. There are multiple series S^k for each DCE-MRI study P_i, where k indicates the index of DCE-MRI series ($k \in \{0, 1, 2..., 5\}$). For one DCE-MRI series S^k, each lesion consists multiple axial slices x_j^k, where j in the range $[j\_min, j\_max]$ ($j\_min = 7, j\_max = 80$). In our case, the mean value of j is 20.6. The dataset $\mathbb{D}$ is redefined as $\mathbb{D} = \{I_{ij}^k\}$, where I_{ij}^k represents the jth lesion slice image extracted from the kth DCE series of the ith patient.

2.2 High-Quality Interpolation of DCE-MRI

Given two DCE-MRI series I_{ij}^{k1} and I_{ij}^{k2} ($k1 < k2$), our goal is to predict the intermediate series $I_{ij}^{k'}$ at series step $k'(k1 < k' < k2)$. For convenience and simplicity, the subscript index ij of I is omitted for single lesion and the superscript k is marked as subscript for both I_{k1} and I_{k2} in the rest of this paper.

A straightforward way for DCE-MRI interpolation is to train a CNN to directly output the intermediate series [15]. However, the network has to estimate both the contrast and appearance agent of lesions. In this way, it is hard to generate high-quality intermediate series and accomplish multi-series interpolation at any series step. Inspired by the video interpolation method [13], the contrast transformations are learned firstly, and then an adaptive warping method is proposed to generate the interpolation series of DCE-MRI.

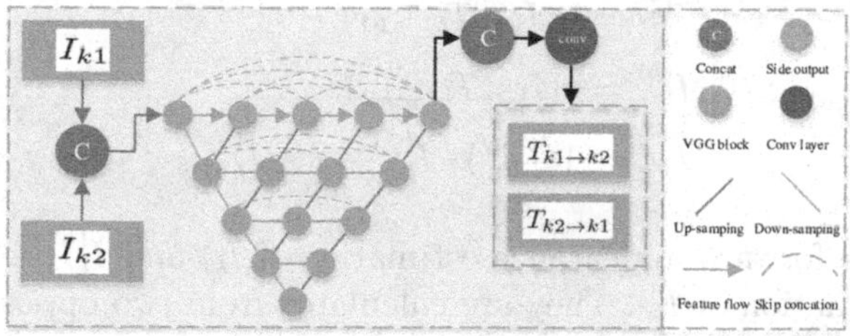

Fig. 2. A contrast transform network based on nested Unet (Unet++) [16] is explored to learn the contrast transformation between the input two DCE-MRI series.

Contrast Transformation Learning: Let $T_{k1\to k2}$ and $T_{k2\to k1}$ denote the kinetics behavior of contrast agent in breast lesions from I_{k1} to I_{k2} and I_{k2} to I_{k1}, respectively. As shown in Fig. 2, the contrast transform network based on nested Unet [16] is used to generate the contrast transform map between I_{k1} and I_{k2}. The nested and dense skip connections of nested Unet have proved effective in recovering fine-gained details of the target objects.

There are total 5 hierarchies of the entire transform network with multiple VGGBlock nodes and the number of filters for the nodes in top-down hierarchies is $[32, 64, 128, 512]$ respectively. Each node is consist of two convolutional layers (kernels 3×3), two batch normalization layers and two ReLU layers. For the nodes in the in-between position of each hierarchy, it is preceded by a concatenation layer that fuses the output from the previous node with the corresponding up-sampled output of the lower node. The Max-pooling layers for down-sampling operation have the same kernel size 2 and stride step 2. Similarly, a bilinear up-sampling operation is utilized to increase the spatial dimension by a factor of 2. The contrast transformations $T_{k1,k2}$ are calculated by extracting the output of the final node, following by a convolutional layer with 4 out channels of size 1×1 convolutional kernel. The former 2 channels of $T_{k1,k2}$ is marked as $T_{k1\to k2}$ and the latter ones is marked as $T_{k2\to k1}$.

Appearance Estimation: The two major modules for appearance estimation are Bi-direction Appearance Estimation (BAE) and Discriminator (D). Inspired by the temporal consistency of video interpolation [13], the intermediate transformations ($T_{k'\to k1}$, $T_{k'\to k2}$) are approximated by combining the bi-directional outputs of the contrast transform network. Assume the temporal length between I_{k1} and I_{k2} is re-scaled to 1, so the range of k' is from 0 to 1.

$$\begin{aligned} T_{k'\to k1} &= -(1-k') \cdot T_{k1\to k2} + k'^2 \cdot T_{k2\to k1} \\ T_{k'\to k2} &= (1-k')^2 \cdot T_{k1\to k2} - k'(1-k') \cdot T_{k2\to k1}. \end{aligned} \tag{1}$$

The intermediate series $I_{k'}$ can be synthesized with a forward appearance estimation and a backward appearance estimation as follows.

$$\begin{aligned} I_{k'}^{(f)} &= g(I_{k1}, T_{k'\to k1}) \\ I_{k'}^{(b)} &= g(I_{k2}, T_{k'\to k2}) \\ \widehat{I}_{k'} &= (1-k') \cdot I_{k'}^{(f)} + k' \cdot I_{k'}^{(b)}. \end{aligned} \tag{2}$$

where $I_{k'}^{(f)}$ is the forward appearance estimation of $I_{k'}$ and $I_{k'}^{(b)}$ is the backward appearance estimation of $I_{k'}$. They are calculated from two opposite directions, using the input image and intermediate contrast transformations. The differentiable function $g(\cdot, \cdot)$ is a backward warping for appearance estimation of $I_{k'}$ [17]. Specially, a mesh-grid is stacked with the intermediate contrast transformations, and then the appearance of $I_{k'}$ can be estimated using a bilinear interpolation with the input image and the mesh-grid. For temporal consistency, the closer the series step k' is to 0, the more contribution the forward appearance estimation makes to $\widehat{I}_{k'}$; a similar property holds for the backward appearance estimation.

The training process for estimating the parameters of contrast transform network is based on the minimization of the sum of mean-square error loss and the loss of Discriminator over the training set. The Discriminator is used to distinguish whether the generated series and the input series belongs to the same DCE-MRI series. It adopts the popular ResNet-50 sub-network for visual feature encoding aiming at distinguishing the interpolation series between the real/fake DCE-MRI series.

3 Experiment

In the experiment, 1223 DCE-MRI slices (each slice with one pre-contrast and five post-contrast) of our dataset are divided into training (70%), test (20%) and validation (10%). All computed breast lesions of a particular patient belong to only one of the three subsets above, ensuring that all methods are not tested using patients seen during the training stage. The best configuration in train-validation setup is used to report the test results.

For training the whole framework, we use SGD to optimize the contrast transform network and Discriminator. Other parameters are as follows: batch size (6), epoch (50), number of training iteration (1000), and learning rate ($10e^{-4}$) with a decay (0.9) for each epoch. We use the Mean Square Error (MSE), Structural SIMilarity (SSIM) to compare the performance of different methods. All experiments are carried out in Python 3.7 using the PyTorch 0.4 libaray, performed under a PC with an Intel CPU with 3.50 GHz and a Tesla K40c GPU.

Table 1. Single-series interpolation performance of DCE-MRI with different methods.

Method	BAE	D	MSE	SSIM
Unet [18]			2.995 ± 1.832	0.722 ± 0.115
Unet++ [16]			2.023 ± 1.395	0.775 ± 0.107
Phase-based [12]			1.025 ± 0.672	0.809 ± 0.083
CCGAN [19]		✓	1.500 ± 1.158	0.829 ± 0.076
Unet	✓		0.771 ± 0.740	0.893 ± 0.060
HQI-DNN	✓		0.763 ± 0.731	0.894 ± 0.061
HQI-DNN	✓	✓	$\mathbf{0.759 \pm 0.730}$	$\mathbf{0.894 \pm 0.060}$

In this section, the single intermediate series interpolation of breast DCE-MRI is performed. We compare our approach with state-of-the-art methods including Unet [18], Unet++ [16], Conditional Convolution GANs (CCGAN) [19] and Phase-based interpolation [12]. The generator of CCGAN shares the same architecture of Unet. The Discriminator of CCGAN has the same architecture with the Discriminator in HQI-DNN. We also investigate the contribution of two critical components, BAE and D.

The average and standard deviations of single-series interpolation performance are summarized in Table 1. Adding each of the two components (BAE

Table 2. The detailed single-series DCE-MRI interpolation performance. I_{k1} and I_{k2} are the start and end series number of DCE-MRI.

Method	$I_{k1} \rightarrow I_{k2}$	MSE	SSIM
Unet [18]	$1 \rightarrow 3$	3.050 ± 1.806	0.719 ± 0.110
	$2 \rightarrow 4$	2.949 ± 1.917	0.724 ± 0.119
	$3 \rightarrow 5$	2.987 ± 1.774	0.722 ± 0.117
Unet++ [16]	$1 \rightarrow 3$	2.135 ± 1.459	0.769 ± 0.107
	$2 \rightarrow 4$	1.783 ± 1.115	0.784 ± 0.104
	$3 \rightarrow 5$	2.151 ± 1.546	0.772 ± 0.111
CCGAN [19]	$1 \rightarrow 3$	1.389 ± 0.935	0.832 ± 0.070
	$2 \rightarrow 4$	1.277 ± 0.740	0.842 ± 0.059
	$3 \rightarrow 5$	1.835 ± 1.560	0.814 ± 0.091
Phase-based [12]	$1 \rightarrow 3$	0.902 ± 0.454	0.808 ± 0.068
	$2 \rightarrow 4$	0.995 ± 0.620	0.816 ± 0.076
	$3 \rightarrow 5$	1.178 ± 0.853	0.802 ± 0.101
HQI-DNN	$1 \rightarrow 3$	0.725 ± 0.671	0.896 ± 0.058
	$2 \rightarrow 4$	$\mathbf{0.613 \pm 0.367}$	$\mathbf{0.904 \pm 0.037}$
	$3 \rightarrow 5$	0.939 ± 0.981	0.882 ± 0.077

and D) can improve interpolation performance. Specifically, the MSE of Unet is greatly decreased to 0.771 when BAE module is imported. The BAE module fuses contrast transformations from two opposite directions, which greatly improves the interpolation quality of in-between series. The discriminator can take a global estimation to weaken the blurry problem and further promote the quality of interpolation image. HQI-DNN with BAE and D performs favorably against all the comparing methods.

The detailed single-series interpolation performance and the interpolating DCE-MRI are shown in Tabel 2 and Fig. 3. Our model consistently outperforms both non-CNN method [12] and CNN-based methods [16,18,19] on each input series pairs. The edge signals and textures features are preserved effectively for both enhanced and un-enhanced regions with HQI-DNN. Besides, there is an interesting phenomenon that almost all methods achieve a better performance on DCE-MRI series $2 \rightarrow 4$ than other input pairs in Table 2. That is mainly because majority of lesions were markedly enhanced in the second post-contrast series of DCE-MRI. Therefore, it is maybe a little easier for DCE-MRI interpolation when the signal of a breast lesion is apparently strengthened. Non-parametric Kruskal-Wallis Test along with Bonferroni correction is employed to compare the image interpolation quality of five algorithms in Table 2. The Kruskal-Wallis test is conducted at the 5% significance level. The median value for MSE and SSIM are 1.192 and 0.834, respectively. Comparing to other four methods, HQI-DNN differs significantly from them ($p < 0.05$).

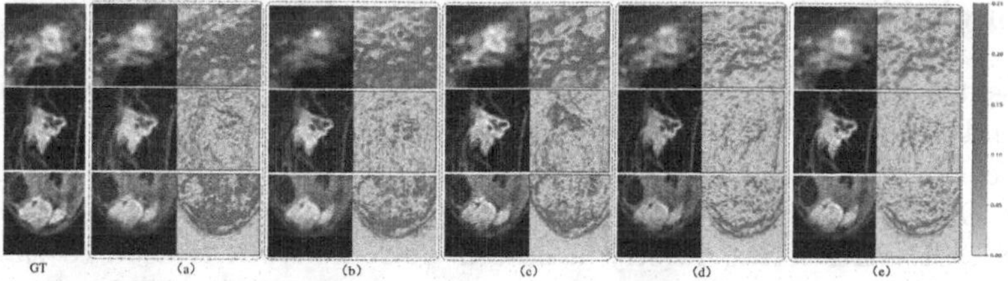

Fig. 3. Visual comparisons. The difference map (right: pcolor) between Ground Truth (GT) and interpolation image (left) is listed. (a) Unet (b) Unet++ (c) CCGAN (d) Phase-based (e) HQI-DNN

The proposed HQI-DNN can produce as many intermediate series as needed between two DCE-MRI series. In this section, we predict three in-between DCE-MRI series. The input series number is $1 \rightarrow 5$ and the intermediate series number for interpolation is 2, 3, and 4. Our method is compared with widely used multi-series interpolation method [12]. As shown in Table 3, HQI-DNN achieves a better performance than Phase-based method [12]. The Phase-based method [12] propagates phase information across oriented multi-scale pyramid level using a novel bounded shift correction strategy. However, some high-pass residual information items are not represented in any pyramid level which incurs some blurring in areas with small motion. In contrast, the nested and dense skip pathways in HQI-DNN have recovered fine-gained details of the targets effectively. Besides, it takes more time for the multi-scale pyramid decomposition in [12] (0.103 s per image) than HQI-DNN (0.016 s per image).

Table 3. Multi-series interpolation performance of DCE-MRI. Evaluation includes MSE, SSIM, and TIME (seconds per DCE-MRI series).

Method	$I_{k1} \rightarrow I_{k2}$	MSE	SSIM	TIME
Phase-based [12]	$1 \rightarrow 5(2)$	1.152 ± 0.610	0.775 ± 0.072	–
	$1 \rightarrow 5(3)$	1.272 ± 0.826	0.757 ± 0.091	–
	$1 \rightarrow 5(4)$	1.401 ± 1.105	0.764 ± 0.111	–
	Average	1.274 ± 1.963	0.765 ± 0.627	0.103
HQI-DNN	$1 \rightarrow 5(2)$	0.996 ± 0.742	0.874 ± 0.046	–
	$1 \rightarrow 5(3)$	1.202 ± 1.040	0.864 ± 0.073	–
	$1 \rightarrow 5(4)$	1.403 ± 1.638	0.852 ± 0.101	–
	Average	$\mathbf{1.200 \pm 2.083}$	$\mathbf{0.863 \pm 0.873}$	**0.016**

4 Conclusion

We present a novel HQI-DNN framework for DCE-MRI interpolation using an end-to-end trainable CNN. It gives a good solution to two key problems of DCE-MRI: inconsistent DCE-MRI series and inadequate image quantity for CAD systems. Our main idea is to learn the kinetic and spatial transformations of different DCE-MRI series. Firstly, a nested CNN network is used to learn the bidirectional contrast transformation between two input DCE-MRI series. Then, the intermediate DCE-MRI series can be synthesized with a pre-defined forward appearance estimation and a backward appearance estimation. Comparing with previous supervised and unsupervised methods, the proposed HQI-DNN is computationally very efficient and simple to implement. In the future, more data should be used to further evaluate our method. We would like to focus on expanding this research on breast cancer diagnosis with the interpolated DCE-MRI.

Acknowledgments. This research was supported by the Key Research and Development Program of Shaanxi Province (the General Project of Social Development) (2020SF-049); Scientific Research Project of Education Department of Shaanxi Provincial Government (19JK0808); Xi'an Science and Technology Plan Project (20YXYJ0010(5)).

References

1. Mihalco, S., Keeling, S., Murphy, S., O'Keeffe, S.: Comparison of the utility of clinical breast examination and MRI in the surveillance of women with a high risk of breast cancer. Clin. Radiol. **75**(3), 194–199 (2020)
2. Luo, L., et al.: Deep angular embedding and feature correlation attention for breast MRI cancer analysis. In: Shen, D., et al. (eds.) MICCAI 2019. LNCS, vol. 11767, pp. 504–512. Springer, Cham (2019). https://doi.org/10.1007/978-3-030-32251-9_55
3. Romeo, V., et al.: Tumor segmentation analysis at different post-contrast time points: a possible source of variability of quantitative DCE-MRI parameters in locally advanced breast cancer. Eur. J. Radiol. **126**, 108907 (2020)
4. Antropova, N., Huynh, B., Giger, M.: Recurrent neural networks for breast lesion classification based on DCE-MRIs. In: Medical Imaging 2018: Computer-Aided Diagnosis, vol. 10575, p. 105752M. International Society for Optics and Photonics (2018)
5. Banaie, M., Soltanian-Zadeh, H., Saligheh-Rad, H.R., Gity, M.: Spatiotemporal features of DCE-MRI for breast cancer diagnosis. Comput. Methods Programs Biomed. **155**, 153–164 (2018)
6. Haarburger, C., et al.: Multi scale curriculum CNN for context-aware breast MRI malignancy classification. In: Shen, D., et al. (eds.) MICCAI 2019. LNCS, vol. 11767, pp. 495–503. Springer, Cham (2019). https://doi.org/10.1007/978-3-030-32251-9_54
7. Marrone, S., Piantadosi, G., Fusco, R., Petrillo, A., Sansone, M., Sansone, C.: An investigation of deep learning for lesions malignancy classification in breast DCE-MRI. In: Battiato, S., Gallo, G., Schettini, R., Stanco, F. (eds.) ICIAP 2017. LNCS, vol. 10485, pp. 479–489. Springer, Cham (2017). https://doi.org/10.1007/978-3-319-68548-9_44

8. El Adoui, M., Mahmoudi, S.A., Larhmam, M.A., Benjelloun, M.: MRI breast tumor segmentation using different encoder and decoder CNN architectures. Computers **8**(3), 52–62 (2019)
9. Huang, P., et al.: CoCa-GAN: common-feature-learning-based context-aware generative adversarial network for glioma grading. In: Shen, D., et al. (eds.) MICCAI 2019. LNCS, vol. 11766, pp. 155–163. Springer, Cham (2019). https://doi.org/10.1007/978-3-030-32248-9_18
10. Cubuk, E.D., Zoph, B., Mane, D., Vasudevan, V., Le, Q.V.: AutoAugment: learning augmentation strategies from data. In: Proceedings of the IEEE Conference on Computer Vision and Pattern Recognition (CVPR), pp. 113–123 (2019)
11. Zhao, A., Balakrishnan, G., Durand, F., Guttag, J.V., Dalca, A.V.: Data augmentation using learned transformations for one-shot medical image segmentation. In: Proceedings of the IEEE Conference on Computer Vision and Pattern Recognition, pp. 8543–8553 (2019)
12. Meyer, S., Wang, O., Zimmer, H., Grosse, M., Sorkine-Hornung, A.: Phase-based frame interpolation for video. In: IEEE Conference on Computer Vision and Pattern Recognition (CVPR) (2015)
13. Jiang, H., Sun, D., Jampani, V., Yang, M.H., Learned-Miller, E., Kautz, J.: Super SloMo: high quality estimation of multiple intermediate frames for video interpolation. In: Proceedings of the IEEE Conference on Computer Vision and Pattern Recognition (CVPR), pp. 9000–9008 (2018)
14. Bao, W., Lai, W.S., Ma, C., Zhang, X., Gao, Z., Yang, M.H.: Depth-aware video frame interpolation. In: Proceedings of the IEEE Conference on Computer Vision and Pattern Recognition (CVPR), pp. 3703–3712 (2019)
15. Long, G., Kneip, L., Alvarez, J.M., Li, H., Zhang, X., Yu, Q.: Learning image matching by simply watching video. In: Leibe, B., Matas, J., Sebe, N., Welling, M. (eds.) ECCV 2016. LNCS, vol. 9910, pp. 434–450. Springer, Cham (2016). https://doi.org/10.1007/978-3-319-46466-4_26
16. Zhou, Z., Rahman Siddiquee, M.M., Tajbakhsh, N., Liang, J.: UNet++: a nested U-Net architecture for medical image segmentation. In: Stoyanov, D., et al. (eds.) DLMIA/ML-CDS -2018. LNCS, vol. 11045, pp. 3–11. Springer, Cham (2018). https://doi.org/10.1007/978-3-030-00889-5_1
17. Liu, Z., Yeh, R., Tang, X., Liu, Y., Agarwala, A.: Video frame synthesis using deep voxel flow. In: IEEE International Conference on Computer Vision (2017)
18. Ronneberger, O., Fischer, P., Brox, T.: U-Net: convolutional networks for biomedical image segmentation. In: Navab, N., Hornegger, J., Wells, W.M., Frangi, A.F. (eds.) MICCAI 2015. LNCS, vol. 9351, pp. 234–241. Springer, Cham (2015). https://doi.org/10.1007/978-3-319-24574-4_28
19. Zhang, H., Cao, X., Xu, L., Qi, L.: Conditional convolution generative adversarial network for bi-ventricle segmentation in cardiac MR images. In: Proceedings of the Third International Symposium on Image Computing and Digital Medicine, pp. 118–122 (2019)

A Method for Tumor Treating Fields Fast Estimation

Reuben R. Shamir[✉] and Zeev Bomzon

Novocure, Haifa, Israel
rshamir@novocure.com

Abstract. Tumor Treating Fields (TTFields) is an FDA approved treatment for specific types of cancer and significantly extends patients' life. The intensity of the TTFields within the tumor was associated with the treatment outcomes: the larger the intensity the longer the patients are likely to survive. Therefore, it was suggested to optimize TTFields transducer array location such that their intensity is maximized. Such optimization requires multiple computations of TTFields in a simulation framework. However, these computations are typically performed using finite element methods or similar approaches that are time consuming. Therefore, only a limited number of transducer array locations can be examined in practice. To overcome this issue, we have developed a method for fast estimation of TTFields intensity. We have designed and implemented a method that inputs a segmentation of the patient's head, a table of tissues' electrical properties and the location of the transducer array. The method outputs a spatial estimation of the TTFields intensity by incorporating a few relevant parameters in a random-forest regressor. The method was evaluated on 10 patients (20 TA layouts) in a leave-one-out framework. The computation time was 1.5 min using the suggested method, and 180–240 min using the commercial simulation. The average error was 0.14 V/cm (SD = 0.06 V/cm) in comparison to the result of the commercial simulation. These results suggest that a fast estimation of TTFields based on a few parameters is feasible. The presented method may facilitate treatment optimization and further extend patients' life.

Keywords: Tumor Treating Fields · Treatment planning · Simulation

1 Introduction

Tumor Treating Fields (TTFields) therapy is an FDA approved treatment for Glioblastoma Multiforme (GBM) and Malignant Pleural Mesothelioma (MPM) [1, 2]. Clinical trials have shown that adding TTFields to standard of care significantly extends Gllioblastoma patient overall survival [1]. Similar improvements were observed in MPM patients [2]. TTFields are delivered non-invasively using pairs of transducer arrays that are placed on the skin in close proximity to the tumor. The arrays are connected to a field generator that when activated generates an alternating electric field in the range of 100–200 kHz that propagates into the cancerous tissue.

N. Burgos et al. (Eds.): SASHIMI 2020, LNCS 12417, pp. 60–67, 2020.
https://doi.org/10.1007/978-3-030-59520-3_7

Recent post-hoc analysis of clinical data showed that delivery of higher field intensities to the tumor is associated with prolonged patient survival [3]. Therefore, placing the transducer arrays such that the TTFields intensity is maximized in the cancerous tissue, has the potential of further extending patients' life.

Finding the array placement that maximizes field intensity in the tumor is an optimization problem that requires calculating the electric field distribution generated by multiple positions of the arrays. Current methods for estimating TTFields intensity distributions rely on finite element methods that are time consuming and may require hours to compute the field generated by a single pair of arrays [4]. Hence, during any practical optimization scheme for TTFields treatment planning, only a limited number of transducer array locations can be evaluated and the optimization result may be suboptimal.

Recent studies have suggested that machine learning methods can be utilized to approximate finite element methods output [4–11]. Benuzzi et al. [8] predicted some mechanical properties in the assembly of a railway axle and wheel. Wang et al. [9] used a regression model to predict road sweeping brush load characteristics. Lostado et al. [11] utilized regression trees to define stress models and later on [10] to determine the maximum load capacity in tapered roller bearings. Guo et al. [12] have proposed to utilize a convolutional neural network for real-time prediction of non-uniform steady laminar flow. Pfeiffer et al. [13] incorporated a convolutional neural network to estimate the spatial displacement of an organ as a response to mechanical forces. Liang et al. [14] have incorporated a deep neural network to estimate biomechanical stress distribution as a fast and accurate surrogate of finite-element methods. Finally, Hennigh et al. [15] have introduced Lat-Net, a method for compressing both the computation time and memory usage of Lattice Boltzmann flow simulations using deep neural networks.

In this study we present a novel method that incorporates the random forest regression for the fast estimation of TTFields. To the best of our knowledge, it is the first attempt to utilize a machine learning method for significantly decreasing the computation time of TTFields simulation. The key contributions of this study are as follows: 1) identification of key parameters that effect TTFields intensity; 2) a method for extraction of these parameters; 3) utilization of random forest regression for fast estimation of the TTFields, and; 4) validation of the method on 10 GBM patients.

2 Methods

2.1 Key Parameters that Effect TTFields

Based on Ohm's law, Maxwell's equations in matter and Coulomb's law, the electric field is inversely related to conductivity (σ), permittivity (ε), and distance from electrical source (d_e), respectively. A close inspection of simulation results (Fig. 1) suggests that the TTFields are larger when the tissue is in the proximity of the cerebrospinal fluid (CSF). A possible explanation for this observation is that electrons are accumulated on the CSF's boundary since of its high conductivity, therefore, increasing the electric potential in these zones. We denote the shortest distance of a voxel from a voxel of CSF as d_c. Another observation is that the TTFields are larger in tissues that are closer to the imaginary line between the centers of TA pairs (Fig. 1). This observation is in line with a generalization of Coulomb's law to finite parallel plates in homogenous matter. We denote the distance between a voxel and the line along TA centers as d_l. The conductivity and permittivity are expected to have a linear relation with the electric field, and the distance is polynomial to the electric field. Yet, we are unfamiliar with a formula that combines all of the above features, and therefore, incorporate a regression method.

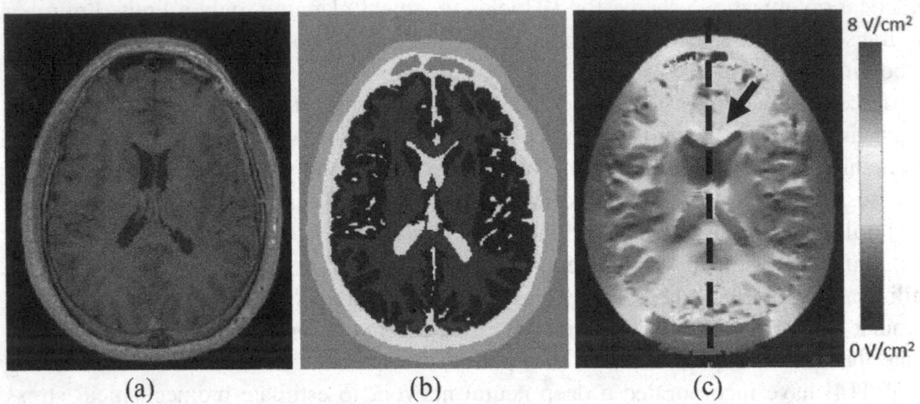

Fig. 1. (a) Head MRI T1 with gadolinium of a GBM patient who underwent TTField treatment. (b) Segmentation of the patient's MRI into tissues with different electrical properties. (c) TTfields spatial distribution that was computed with a finite element methods. Note that the TTFields are increased in the vicinity of cerebrospinal fluid (arrow). Moreover, TTFields are larger in tissues that are closer to the dashed line between the centers of TA pairs.

Given patient's head MRI the above key parameters were extracted as follows. At first, we have segmented the head into eight tissues (Fig. 1b): 1) skin and muscle (as one tissue); 2) skull; 3) CSF; 4) white matter; 5) grey matter; 6) tumor – enhancing; 7) tumor – necrotic, and; 8) tumor resection cavity. The segmentation of the tumor was performed semi-automatically using region growing and active contours methods [6]. The segmentation of the head tissues (1–5) was performed automatically with a custom atlas-based method [16]. The conductivity and permittivity of the different tissues were determined as described in [17]. The distances of each voxel from electrical source, CSF and the line along TA centers were efficiently computed using the method presented in Danielsson et al. [5].

2.2 Random Forests Regression for Estimation of TTFields

Random forests are an ensemble of decision tree predictors, such that each tree is restricted by a random vector that governs the sensitivity of the tree to the input features [18]. Lostado et al. [11] have demonstrated that regression trees facilitate effective modeling of FEM-based non-linear maps for fields of mechanical force.

Moreover, they suggest that since random forests divide the dataset into groups of similar features and facilitate local group fitting, good models can be generated also when the data is heterogeneous, irregular, and of limited size.

Therefore, we have incorporated a random forest regressor. We set up 30 trees, mean squared error quality of split measure, using bootstrap and out-of-bag samples to estimate regression quality on unseen samples. The number of trees was selected by a trial-and-error process to balance accuracy and prediction-time tradeoff. The input per voxel to the regression tree is as follows. 1) conductivity (σ); 3) permittivity (ε); 4) distance from closest electrical source (d_e); 5) distance from closest CSF (d_c), and; 6) distance from TAs midline (d_l).

We investigated the relevance of the above features to the prediction in our experimental setup (see below) using mean decrease in impurity method that results with a feature's importance score in the range of 0 to 1 [18]. The distance of closest electrical source was by far the most important feature (0.65). Distance from TA midline and from CSF were of secondary importance (0.15 and 0.1, respectively). Conductivity and permittivity importance scores were both 0.05. A typical example for a decision tree in the random forest is presented in Fig. 2. An interesting observation is that the mean squared error (MSE) was reduced for locations that are further from the TA. The specific tree in this example splits the data at 16.6 mm distance to TA. Indeed, larger errors were observed for the 15% of data that are within this range.

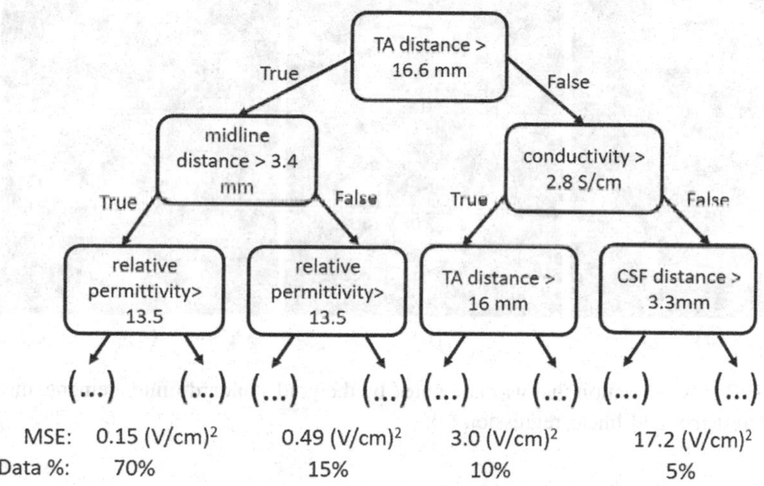

Fig. 2. The first three layers of a decision tree regressor that was trained to predict the tumor treating fields strength. The mean squared error (MSE) was reduced for locations that are further from the transducer array (TA): compare the left and right branches' accuracies in this example.

In addition, we compared the random forest to a multi-linear regression. Specifically, the following linear formula was incorporated to estimate the TTFields.

$$|E| \sim a_0 + a_1\sigma^{-1} + a_2\varepsilon^{-1} + a_3d_e^{-1} + a_4d_e^{-2} + a_5d_c + a_6d_l \quad (1)$$

The coefficients a_i were computed to best fit the finite elements method output to the linear regression model (see next section).

2.3 Experimental Setup

We have validated the suggested method using a dataset of 10 patients that underwent TTFields therapy. At first, the patients' MRIs were segmented as described in Sect. 2.1 and the head's outer surface was extracted using the marching cubes algorithm [19]. Then, two TA pairs were virtually placed on the head's surface. The first pair was placed such that one TA is on the forehead and the other one is on the back of the head. In this case, the TTField direction is roughly parallel to the anterior-posterior (AP) axis. The second pair was placed such that the TAs are on opposite lateral sides of the head. In this case, the TTField direction is roughly parallel to the left-right (LR) axis of the head. For each of the 20 pairs, we computed the absolute electric field intensity spatial distribution with a finite element method (Fig. 3). That is, we associate the electric field for each voxel in the patient's MRI. We marked this dataset as gold standard as it was verified in phantoms and associated with patients' survival [3].

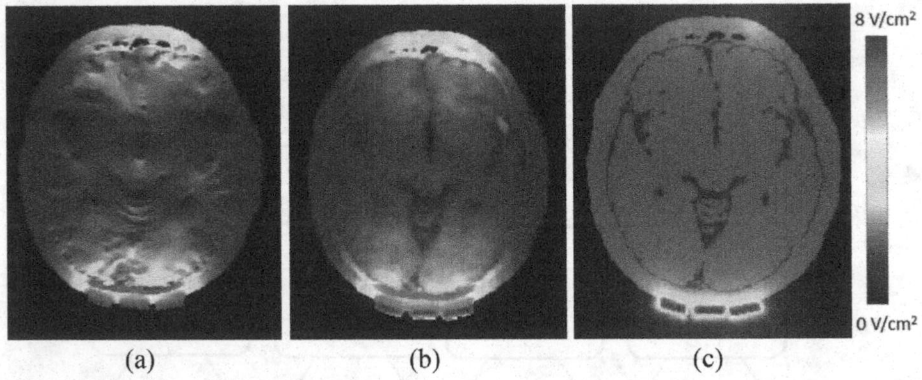

Fig. 3. TTFields estimation that was computed by the gold standard finite-elements method (a), random forest (b), and linear regression (c).

We use a leave-one-out approach for the training. One test patient was excluded at a time while the 18 datasets of the rest nine patients were incorporated to train the random forest and the multilinear regression model (Eq. 1). Then, the TTFields were predicted using random forest and multilinear regression model on the test patient data. Large parts of the image are associated with air that is not conductive. Therefore, it can bias the result of the model. To handle this situation, we consider only a small portion of the voxels with air in the training by ensuring that their number is similar to those in other segmented tissues. In this study, the training and prediction are performed per voxel independently of its neighbors.

We have implemented the suggested method with Python 3.6 using scipy [20], numpy [21], scikit-learn [22] and SimpleITK [23] packages. We used 3D Slicer [24] for visual inspection of the results. The method was executed on a standard desktop computer (Intel i7 CPU, 16 GB RAM) with Windows 10 operating system (Microsoft, Redmond, WA, USA). The gold standard simulations were computed using sim4life (Zurich Med Tech, Zurich, Switzerland) on a dedicated simulation computer (Intel i7 CPU, NVidia 1080 Ti GPU, 128 GB RAM) with Windows 10 operating system (Microsoft, Redmond, WA, USA). Patients' MRIs were T1 weighted with gadolinium with voxel spacing of $1 \times 1 \times 1$ mm^3 and incorporated the entire head.

3 Results

The average absolute differences between the random forest prediction and the gold standard was 0.14 V/cm (patients' SD = 0.035, range 0.08 – 0.23 V/cm, N = 20). Table 1 presents a per-patient summary of our results. The random forest resulted with a better accuracy in comparison to the multilinear regression. Compare the random forest average absolute differences above and the linear regression results of 0.29 V/cm (patients' SD = 0.04, range 0.23 – 0.37 V/cm, N = 20). The random forest average prediction time was 15 s (SD = 1.5 s). Note that this measure is excluding the pre-processing required to extract the distance measures that typically required 30 s for each: TA, CSF and midline.

Table 1. Experiment results. The average (SD) of absolute differences between the TTFields computed with a finite element method and the random forest. Results for pairs of transducer arrays along the anterior-posterior (AP) and left-right (LR) axes of the head are presented.

Patient #	AP error (V/cm)	LR error (V/cm)	Patient #	AP error (V/cm)	LR error (V/cm)
1	0.14 (0.62)	0.23 (0.80)	6	0.08 (0.45)	0.08 (0.43)
2	0.16 (0.56)	0.19 (0.71)	7	0.15 (0.61)	0.14 (0.62)
3	0.15 (0.69)	0.17 (0.59)	8	0.10 (0.50)	0.11 (0.49)
4	0.15 (0.61)	0.18 (0.66)	9	0.14 (0.60)	0.14 (0.64)
5	0.15 (0.61)	0.15 (0.59)	10	0.11 (0.55)	0.13 (0.62)

Figure 3 presents typical electric-field spatial distributions that were computed by the gold-standard, the random forest, and the multilinear regression. Figure 4 demonstrates typical absolute differences between the gold-standard and random-forest prediction. The values are very similar (<0.4 V/cm) in most locations. However, large errors (>2 V/cm) were observed in the vicinity of the TAs (Fig. 4a) and instantly outside the ventricles along the TA main axis (Fig. 4b).

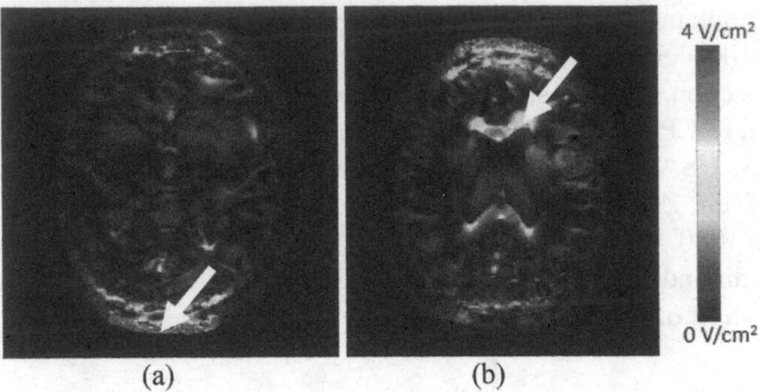

Fig. 4. Absolute differences between gold standard finite-elements method and random-forest based estimation. Larger errors were observed near the transducer arrays (a) and near the ventricles along the main transducer array axis (b)

4 Discussion

We have presented a novel method for the fast estimation of TTFields spatial distribution and demonstrated that average accuracy of 0.14 V/cm can be achieved within a short time. Compare the ~ 1.5 min computation time with the suggested random forest method to the 3–4 h computation time using the gold standard method. Note that the computation time can be further reduced by a factor of three by the parallelization of data preparation.

Selection of optimal TA location involves the computation of average TTFields over a tumor area. Averaging is expected to further improve accuracy. Yet, the utilization of the random forest method TTFields estimation for optimization of TA placement is out of the scope of this study and requires further investigation.

One limitation of our method is that it assumes that the effect of neighbor voxels is minor and can be neglected. We plan to revise our method to incorporate also neighbor voxels and consider convolutional and recurrent neural networks. Another limitation is the data-preparation computation time. We plan to investigate one-time distance-maps computation and fast manipulation of these maps upon alternation of TA locations to reduce overall computation time to a few seconds. Last, we plan to extend the method to the chest and abdomen for supporting additional indications of TTFields treatment.

References

1. Stupp, R., Taillibert, S., et al.: Effect of tumor-treating fields plus maintenance temozolomide vs maintenance temozolomide alone on survival in patients with glioblastoma: a randomized clinical trial. JAMA **318**, 2306–2316 (2017)
2. Ceresoli, G., Aerts, J., et al.: Tumor Treating Fields plus chemotherapy for first-line treatment of malignant pleural mesothelioma. Int. J. Radiat. Oncol. **104**, 230–231 (2019)
3. Ballo, M.T., et al.: Correlation of tumor treating fields dosimetry to survival outcomes in newly diagnosed glioblastoma. Int. J. Radiat. Oncol. Biol. Phys. **104**, 1106–1113 (2019)

4. Wenger, C., et al.: A review on tumor-treating fields (TTFields): clinical implications inferred from computational modeling. IEEE Rev. Biomed. Eng. **11**, 195–207 (2018)
5. Danielsson, P.E.: Euclidean distance mapping. Comput Graph Img Proc. **14**, 227–248 (1980)
6. Caselles, V., et al.: Geodesic active contours. Int. J. Comput. Vis. **22**, 61–79 (1997)
7. Koessler, L., et al.: In-vivo measurements of human brain tissue conductivity using focal electrical current injection through intracerebral multicontact electrodes. Hum. Brain Mapp. **38**, 974–986 (2017)
8. Benuzzi, D., et al.: Prediction of the press-fit curve in the assembly of a railway axle and wheel. Proc. Inst. Mech. Eng. Part F J. Rail Rapid Transit. **218**, 51–65 (2004)
9. Wang, C., et al.: Regression modeling and prediction of road sweeping brush load characteristics from finite element analysis and experimental results. Waste Manage. **43**, 19–27 (2015)
10. Lostado-Lorza, R., et al.: Using the finite element method and data mining techniques as an alternative method to determine the maximum load capacity in tapered roller bearings. J. Appl. Log. **24**, 4–14 (2017)
11. Lostado, R., et al.: Combining regression trees and the finite element method to define stress models of highly non-linear mechanical systems. J. Strain Anal. Eng. Des. **44**, 491–502 (2009)
12. Guo, X., Li, W., Iorio, F.: Convolutional neural networks for steady flow approximation. In: Proceedings of the ACM SIGKDD International Conference on Knowledge Discovery and Data Mining, pp. 481–490. Association for Computing Machinery (2016)
13. Pfeiffer, M., Riediger, C., Weitz, J., Speidel, S.: Learning soft tissue behavior of organs for surgical navigation with convolutional neural networks. Int. J. Comput. Assist. Radiol. Surg. **14**(7), 1147–1155 (2019). https://doi.org/10.1007/s11548-019-01965-7
14. Liang, L., Liu, M., et al.: A deep learning approach to estimate stress distribution: a fast and accurate surrogate of finite-element analysis. J. R. Soc. Interface. **15** (2018)
15. Hennigh, O.: Lat-Net: Compressing Lattice Boltzmann Flow Simulations using Deep Neural Networks (2017)
16. Bomzon, Z., et al.: Using computational phantoms to improve delivery of Tumor Treating Fields (TTFields) to patients. In: Proceedings of the Annual International Conference of the IEEE Engineering in Medicine and Biology Society, EMBS, pp. 6461–6464 (2016)
17. Bomzon, Z., Hershkovich, H.S., et al.: Using computational phantoms to improve delivery of Tumor Treating Fields (TTFields) to patients. In: 2016 38th Annual International Conference of the IEEE Engineering in Medicine and Biology Society (EMBC), pp. 6461–6464. IEEE (2016)
18. Breiman, L.: Random forests. Mach. Learn. **45**, 5–32 (2001)
19. Lorensen, W.E., Cline, H.E.: Marching cubes: a high resolution 3D surface construction algorithm. SIGGRAPH **1987**, 163–169 (1987)
20. Jones, E., et al.: SciPy: Open source scientific tools for Python. http://www.scipy.org/
21. Oliphant, T.E.: A Guide to NumPy. Trelgol Publishing, USA (2006)
22. Pedregosa, F., Varoquaux, G., et al.: Scikit-learn: machine learning in Python. J. Mach. Learn. Res. **12**, 2825–2830 (2011)
23. Yaniv, Z., Lowekamp, B.C., Johnson, H.J., Beare, R.: SimpleITK image-analysis notebooks: a collaborative environment for education and reproducible research. J. Digit. Imaging **31**(3), 290–303 (2017). https://doi.org/10.1007/s10278-017-0037-8
24. Fedorov, A., Beichel, R., et al.: 3D Slicer as an image computing platform for the Quantitative Imaging Network. Magn. Reson. Imaging **30**, 1323–1341 (2012)
25. Goodfellow, I., Bengio, Y., Courville, A.: Deep Learning. MIT Press, Cambridge (2016)

Heterogeneous Virtual Population of Simulated CMR Images for Improving the Generalization of Cardiac Segmentation Algorithms

Yasmina Al Khalil[1(✉)], Sina Amirrajab[1], Cristian Lorenz[2], Jürgen Weese[2], and Marcel Breeuwer[1,3]

[1] Eindhoven University of Technology, Eindhoven, The Netherlands
{y.al.khalil,s.amirrajab,m.breeuwer}@tue.nl

[2] Philips Research Laboratories, Hamburg, Germany
{cristian.lorenz,juergen.weese,}@philips.com

[3] Philips Healthcare, MR R&D - Clinical Science, Best, The Netherlands

Abstract. Simulating a large set of medical images with variability in anatomical representation and image appearance has the potential to provide solutions for addressing the scarcity of properly annotated data in medical image analysis research. However, due to the complexity of modeling the imaging procedure and lack of accuracy and flexibility in anatomical models, available solutions in this area are limited. In this paper, we investigate the feasibility of simulating diversified cardiac magnetic resonance (CMR) images on virtual male and female subjects of the eXtended Cardiac and Torso phantoms (XCAT) with variable anatomical representation. Taking advantage of the flexibility of the XCAT phantoms, we create virtual subjects comprising different body sizes, heart volumes, and orientations to account for natural variability among patients. To resemble inherent image quality and contrast variability in data, we vary acquisition parameters together with MR tissue properties to simulate diverse-looking images. The database includes 3240 CMR images of 30 male and 30 female subjects. To assess the usefulness of such data, we train a segmentation model with the simulated images and fine-tune it on a small subset of real data. Our experiment results show that we can reduce the number of real data by almost 80% while retaining the accuracy of the prediction using models pre-trained on simulated images, as well as achieve a better performance in terms of generalization to varying contrast. Thus, our simulated database serves as a promising solution to address the current challenges in medical imaging and could aid the inclusion of automated solutions in clinical routines.

Keywords: Image simulation · Transfer learning · Virtual population · Cardiac magnetic resonance imaging · Cardiac segmentation

---

Y. Al Khalil and S. Amirrajab—Contributed equally.

N. Burgos et al. (Eds.): SASHIMI 2020, LNCS 12417, pp. 68–79, 2020.
https://doi.org/10.1007/978-3-030-59520-3_8

1 Introduction

Being one of the most valuable sources of diagnostic information, there is a notable demand for fast and precise analysis of medical images, which is still largely dependent on human interpretation. Recent advances in deep learning (DL) can address this problem by automation. However, DL-based methods are still not directly employed in real-world applications, as they typically rely on the availability of accurately annotated data. Annotating medical images is a time consuming process that requires expert supervision and is limited by data protection and patient confidentiality [13]. Moreover, significant variations in contrast, resolution, signal-to-noise ratio and image quality are often observed in images collected from various clinical sites and machines, causing a deterioration in the performance of DL-based methods [10]. In theory, robustness could be increased by introducing highly heterogeneous data-sets to the training and testing procedure, representative of the data the network will likely see in the future. However, this approach implies the collection and annotation of a significantly large data-set covering all possible cases, which is not scalable to real-world applications and is limited by data protection measures.

In recent years, physics-based image simulation as well as data-driven image synthesis have addressed the lack of properly annotated, diverse data suitable for developing DL algorithms [6]. For the application of cardiac MR (CMR) image simulation, MRXCAT approach [14] grounded on the extended Cardiac-Torso (XCAT) anatomical phantom [11] has demonstrated a great utility for optimizing advanced acquisition and reconstruction methods in CMR studies. However, the usefulness of such simulated data for generating a huge number of images for DL-based image analysis tasks remains very limited due to insufficient image realism, lack of variability, low quality, and quantity of images. Novel image synthesis approaches based on generative adversarial networks (GANs) [7] have been developed very recently to generate realistic-looking images suitable for alleviating data scarcity and generalization problems [15]. Unpaired image-to-image translation to synthesize CMR images from CT images [5] and factorized representation learning to synthesize controllable 3D images [9] are among very recent advances. Although these models can generate high-quality images, the ground truth anatomical representation of the images is limited to the training data and is not necessarily plausible in terms of underlying organ and structure. Therefore, such data may have limited application to the downstream supervised tasks, where the accuracy of the ground truth labels is essential. To address this issue, recent work by [1] combines the anatomical information derived from a physics-based anatomical model with the data-driven imaging features learned from real data in a conditional setting. This hybrid method achieves more control over both anatomical content and the overall style of the images. However, image synthesis with varying anatomies, local contrast, and noise is still lacking.

In this paper, we propose a physics-based simulation strategy to generate a large and diverse virtual population of CMR images[1] including virtual subjects

[1] The simulated database will be available online for medical imaging research.

with both anatomical and contrast variations. Benefiting from the flexibility of the XCAT anatomical model in creating computerized models with user-defined parameters, we create adult male and female subjects with variations in their anatomical representation. This resembles the real differences between different patients in terms of body sizes and heart features. We assign unique MR tissue properties to each individual subject. The signal intensity of the image is governed by the formula of cine CMR imaging based on the solution of Bloch equations for steady-state free precession sequence. The contrast variation is the result of combining tissue properties with acquisition parameters in our physics-based CMR image simulation pipeline which is an improved version of the MRXCAT approach. This combination introduces a great variation in the image contrast and appearance, providing enough diversity in data to train a segmentation model agnostic to inherent contrast variability in the real data. We hypothesize that such a heterogeneous population can aid in developing a deep learning-based algorithm more robust and generalizable to unseen data with unknown image characteristics. To explore whether this is truly the case, we utilize a transfer learning approach for training a segmentation model for the task of heart cavity segmentation in short-axis CMR images. Our model is pre-trained only on the data obtained through the proposed simulation strategy and fine-tuned on a small data-set of real CMR data. By deploying such a model on the data coming from various sites and scanners, we compare its generalization and adaptation capability to models trained only on real data, particularly when such data is scarce.

2 Methodology

2.1 CMR Image Simulation Approach

The CMR image simulation is grounded on the Bloch equations for cine contrast. We utilize the MRXCAT numerical framework as the basis of our approach to simulate a diverse image database on the XCAT anatomical phantoms. We improve the MRXCAT that yields more realistic-looking images with controllable imaging parameters such as SNR, TR, TE, and flip angle as well as MR tissue properties such as T1, T2, and proton density. We consider these defined parameters as an input variable such that we can alter to generate images with variable noise levels and image appearances. To introduce anatomical variations in the database, we employ the parameterized model of the XCAT phantoms.

2.2 Generation of a Diverse Virtual Population

The overview of our strategy to simulate a diverse virtual population of CMR images with anatomical and contrast variability is shown in Fig. 1. To make the database heterogeneous, we change three categories of parameters: i) anatomical parameters, ii) MR tissue properties and iii) imaging parameters. To create virtual subjects with variable anatomical representation, we utilize the XCAT

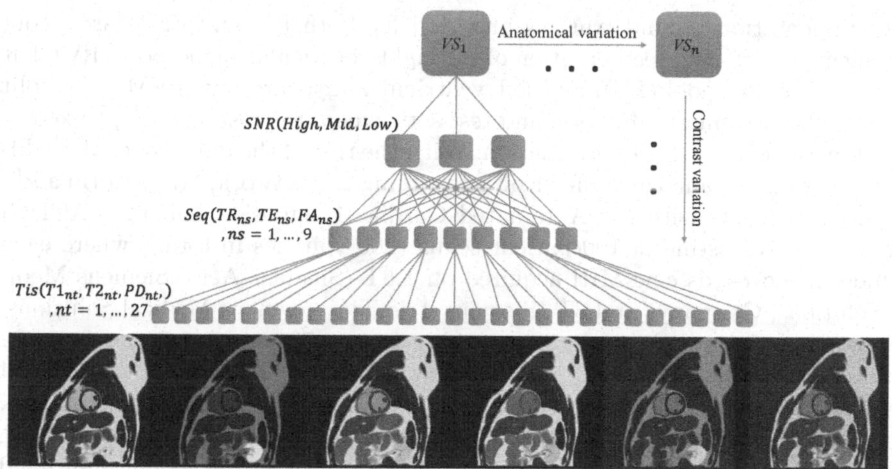

Fig. 1. Simulation strategy to generate a diverse virtual population. Given one virtual subject (VS) with specific anatomical characteristics, we simulate CMR images with varying SNR levels (high, medium, low), sequence parameters (TR, TE, Flip Angle), and tissue properties (T1, T2, Proton Density). The combination of these parameters results in 27 distinct image appearances. The last row shows 6 samples of simulated images for one VS with varying contrast for high to low SNR levels (left to right).

phantoms for normal male and female anatomical models and change the body size in anterior-posterior and lateral directions, the volume of the heart at end-diastolic (ED) and end-systolic (ES) phases, and its orientation and location with respect to surrounding organs. We vary T1 and T2 relaxation times and proton density for more than 12 tissues visible in the field of view. We slightly alter repetition time (TR), echo time (TE) and flip angle for balance steady-stare free precession sequence and change the noise standard deviation to achieve three levels of signal-to-noise-ratios. Given the reported mean and standard deviation in literature [4,12] for the tissue properties, we generate new unique tissue values for each simulation. In the proposed setting, we consider three levels of variations (SNR, sequence, and tissue) for which we have three parameters to change. This results in 27 distinct variations in contrast per subject. In total, we simulate 3240 CMR images including 30 female and 30 male anatomical variations, 27 contrast variations for each subject, and 2 heart phases - ED and ES.

2.3 Data

We utilize one simulated short-axis CMR image data-set and two real short-axis CMR image data-sets in the proposed work. The simulated data-set is generated as described in Sect. 2.1, from which we choose 2500 image volumes for pre-training the model. For fine-tuning, we utilize 183 internal clinical CMR (cCMR) volumes, obtained from six different sites, with highly heterogeneous contrasts due to differences in scanner vendors and models, as summarized in Table 1.

The segmentation ground truth is provided for both ED and ES phases, containing expert manual segmentation of the right ventricular blood pool (RV), left ventricular blood pool (LV), and left ventricular myocardium (LVM). We split the data into training, validation and test sets, where the test set is kept fixed at 80 volumes and set aside for evaluation. With the rest of the data, we explore different subset sizes and their effects on fine-tuning the network. We perform additional experiments with the Automated Cardiac Diagnosis Challenge (ACDC) data-set [3], consisting of 100 patient scans (200 volumes in total), where each includes a short-axis cine-MRI acquired on 1.5T (Siemens Aera, Siemens Medical Solutions, Germany) and 3T (Siemens Trio Tim, Siemens Medical Solutions, Germany) systems with resolutions ranging from 0.70 mm × 0.70 mm to 1.92 mm × 1.92 mm in-plane and 5 mm to 10 mm through-plane. RV, LV and LVM segmentation masks are provided at both ED and ES phases of each patient. We set aside 80 volumes from the ACDC data-set for testing, and use it to both evaluate the model fine-tuned on cCMR data, as well as the model fine-tuned on the rest of the available ACDC volumes.

Table 1. Details of the cCMR data-set used for fine-tuning the models.

Dataset	Total N	Train N	Test N	MRI scanner attributes		Image spatial resolution	
				Manufacturer	Magnetic field strength	In-plane resolution	Slice thickness
Site A	10	5	5	Philips Ingenia	1.5 T	1.8 mm^2/pixel	8 mm
Site B	22	12	10	Philips Achieva	1.5 T	1.8 mm^2/pixel	8 mm
Site C	5	0	5	Philips Intera	1.5 T	2.1 mm^2/pixel	8 mm
Site D	110	80	30	Philips Ingenuity	3.0 T	1.56 mm^2/pixel	8 mm
Site E	12	6	6	Philips Ingenia	1.5 T	1.56 mm^2/pixel	8 mm
Site F	24	0	24	Siemens Aera	1.5 T	2.1–2.78 mm^2/pixel	8 mm

2.4 Segmentation Approach

A 3D U-Net architecture is chosen for both the networks trained from scratch and the pre-trained model, with the addition of batch normalization, dropout (dropout rate of 0.5) and leaky ReLU activation. A common strategy that is shown to significantly improve the training of DL models from scratch is the random initialization of model weights, typically done by Gaussian random initialization. However, methods such as Xavier and He initialization are shown to more effectively avoid the problem of vanishing and exploding gradients during training [8]. We choose to initialize our networks using the He initialization, as it works well with ReLU-based activation functions. Networks are trained using a Focal-Tversky loss [2], designed specifically to address the issue of data imbalance in medical imaging. Even though standard approaches in the literature use the sum of Dice and cross-entropy loss for training, our experiments have shown a consistent improvement in performance when utilizing the Focal-Tversky loss. We use Adam for optimization, with an initial learning rate of $5 * 10^{-3}$, where $\beta_1 = 0.9$, $\beta_2 = 0.999$ and $\eta = 1 * e^{-8}$ and a weight decay of $3 * e^{-5}$. During training, the learning rate is reduced by 0.1 if no improvement is seen in the validation loss for 50 epochs. We utilize two models for training, adjusted based on the training data, ACDC and cCMR data-sets, respectively. Architecturally, the models differ in depth, where the network proposed for training on the cCMR data-set consists of six layers, while the model utilized for ACDC data uses five layers. Both models are trained with a batch size of 4, with cCMR and ACDC data resampled to a median voxel spacing of 1.25 mm × 1.25 mm × 8 mm and 1.5625 mm × 1.5625 mm × 10 mm and rescaled to patch sizes of 320 × 320 × 16 and 224 × 256 × 10, respectively. To increase robustness, we apply random flipping, rotation by integer multiples of $\pi/2$, random Gaussian noise, transposing, random scaling and random elastic deformations during training. All models have converged in 250 epochs.

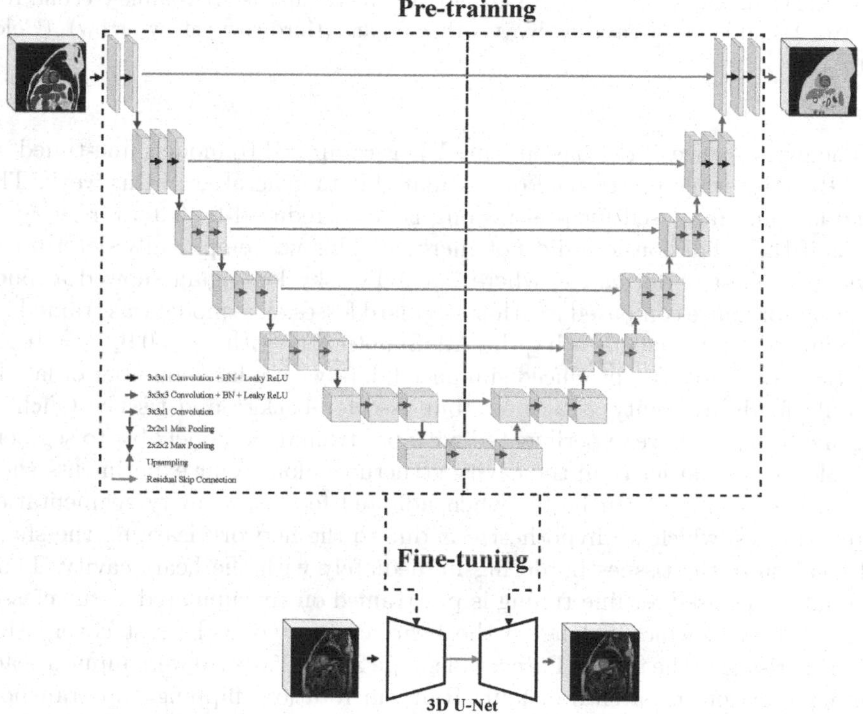

Fig. 2. Pre-training and fine-tuning a 3D U-Net model using simulated images for multi-tissue segmentation. Pre-trained weights are transferred to another model for fine-tuning on a varying number of real MR images for heart cavity segmentation.

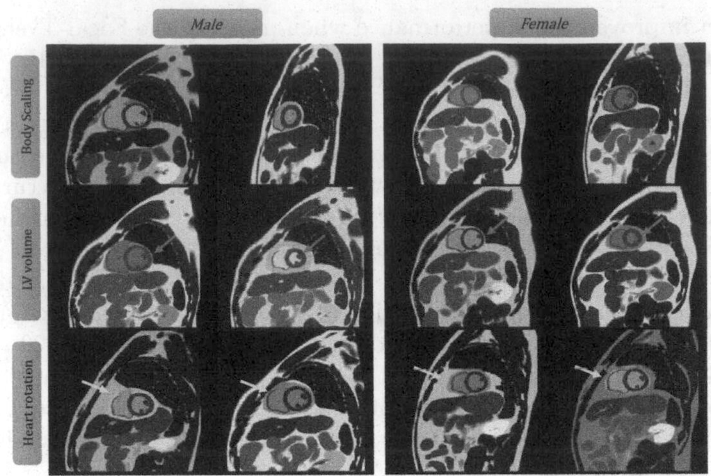

Fig. 3. Anatomical variability of the simulated CMR images for male and female anatomies including body scaling (first row), left-ventricular heart volume (second row red arrow), and heart orientation w.r.t. other organs (third row yellow arrow). (Color figure online)

The performance of the baseline models is compared to models fine-tuned on real MRI data while pre-trained on simulated data generated in this work. The initial learning rate used for pre-training is 10^{-4}, reduced by 0.9 after every 25 epochs if the validation loss did not increase. This was empirically determined to yield the best performance, where Focal-Tversky loss again showed a much better performance compared to other standard losses. The model was trained for a maximum of 1000 epochs using the Adam optimizer with $\beta_1 = 0.9$, $\beta_2 = 0.999$ and $\eta = 1 * e^{-8}$. Having produced simulated data with a large number of labels, not only for heart cavity structures, but also for background tissues (such as lung and liver), we have experimented with pre-training a model able to segment multiple tissues, rather than the cavity structures alone. Our experiments show that such models perform better when adapted for heart cavity segmentation on real images, which we hypothesize is due to the network learning the shape and location of the tissues bordering immediately with the heart cavity. Thus, our final model used for fine-tuning is pre-trained on six simulated tissue classes in total, three of which belong to the heart cavity, while the rest cover other regions of tissue in the image. Patches of shape $256 \times 256 \times 20$ were input at each training iteration in batches of four. Random rotation, flipping and transpose operation were applied as data augmentation. A detailed model architecture is shown in Fig. 2. We reduce the learning rate by half and fine-tune all model layers on the target real MR data. While some studies report that fine-tuning only a part of network layers results in better performance, our experiments suggest that fine-tuning the entire set of pre-trained weights led to models that perform better than or as good as model pre-trained partially. The exponential moving average of the training loss is used as an indicator of whether the learning rate should be reduced, where we take into account the last 30 epochs and reduce it by a factor of 0.2. We further apply early-stopping on the validation set to avoid overfitting. We apply the same approach to data augmentation as in the networks trained from scratch. All fine-tuned models converge in less than 150 epochs.

2.5 Experiments and Results

CMR image simulated cases with anatomical variability including body scaling factor, left-ventricular heart volume, and heart orientation for male and female anatomical models are shown in Fig. 3. Note that the appearance of the simulated images is also changing due to different combinations of acquisition parameters and tissue properties which are unique for each image. We quantitatively evaluate the usefulness of the simulated images by utilizing them for pre-training a segmentation model, fine-tuned on real CMR data and comparing the achieved performance with a model trained from scratch only on real CMR data. In the first set of experiments, we evaluate the effect of model generalization to data acquired from different sites and scanners. Then, we gradually reduce the number of real CMR volumes used for fine-tuning, with the aim to assess if the pre-trained model on simulated data can aid smaller data-sets available for training, while retaining the same performance compared to training with the full set of real images. We use the same subsets of ACDC and cCMR data for inference in all experiments. We run each model for 10 times on all target tasks and report the average Dice scores. Note that site C and site F are excluded from the train set and used for inference only. These data-sets were of particular interest for the problem of generalization due to quite significant variations in resolution, scan orientation and contrast. As described in Table 1, 103 volumes are utilized for training the baseline cCMR segmentation network from scratch, as well for fine-tuning. The same train set is then gradually reduced to evaluate the impact of pre-training with simulated data on model segmentation performance. Detailed configuration summary and results of experiments can be found in Table 2.

Table 2. Segmentation results of the models trained from scratch and fine-tuned on cCMR dataset. All metrics are average Dice scores obtained for each model.

Train N	Pre-training with simulated data	cCMR Test Set: Site A (N=5)			Site B (N=10)			Site C (N=5)			Site D (N=30)			Site E (N=6)			Site F (N=24)			ACDC Test Set N=80		
		LV	RV	LVM	LV	RV	LVM	LV	RV	LVM	LV	RV	LVM	LV	RV	LVM	LV	RV	LVM	LV	RV	LVM
103	NO	**0.956**	0.928	0.890	0.954	0.930	0.899	**0.948**	0.918	0.846	0.934	**0.902**	**0.897**	**0.962**	0.947	0.906	0.881	0.794	0.769	0.858	0.841	0.801
	YES	0.954	**0.941**	**0.896**	0.954	**0.947**	**0.901**	0.945	**0.928**	**0.852**	0.934	0.901	0.889	0.958	**0.952**	0.907	**0.892**	**0.821**	**0.774**	**0.885**	**0.862**	**0.834**
83	NO	0.950	0.916	0.869	0.937	0.913	0.881	0.921	0.894	0.821	0.933	0.897	0.843	0.941	0.941	0.869	0.864	0.782	0.731	0.852	0.823	0.796
	YES	**0.953**	**0.933**	**0.894**	**0.949**	**0.928**	**0.899**	**0.941**	**0.905**	**0.838**	0.933	**0.901**	**0.881**	**0.944**	**0.950**	**0.881**	**0.889**	**0.815**	**0.769**	**0.881**	**0.844**	**0.819**
63	NO	0.944	0.905	0.856	0.921	0.897	0.864	0.920	0.884	0.815	0.928	0.888	0.841	0.933	0.929	0.854	0.853	0.768	0.721	0.848	0.817	0.785
	YES	**0.952**	**0.918**	**0.884**	**0.929**	**0.901**	**0.882**	**0.934**	**0.893**	**0.825**	0.929	**0.892**	**0.854**	0.933	**0.943**	**0.862**	**0.871**	**0.802**	**0.758**	**0.876**	**0.840**	**0.811**
43	NO	0.927	0.887	0.833	0.920	0.885	0.861	0.912	0.877	0.801	0.919	0.876	0.832	0.929	0.911	0.846	0.822	0.721	0.716	0.841	0.792	0.751
	YES	**0.952**	**0.915**	**0.878**	0.921	**0.897**	**0.880**	**0.931**	**0.889**	**0.820**	**0.924**	**0.893**	**0.851**	0.928	**0.932**	**0.857**	**0.869**	**0.791**	**0.747**	**0.871**	**0.832**	**0.803**
23	NO	0.893	0.869	0.822	0.887	0.861	0.849	0.895	0.860	0.778	0.905	0.843	0.812	0.909	0.886	0.818	0.809	0.691	0.672	0.833	0.788	0.746
	YES	**0.942**	**0.913**	**0.869**	**0.910**	**0.886**	**0.876**	**0.921**	**0.885**	**0.817**	**0.921**	**0.884**	**0.847**	**0.921**	**0.915**	**0.848**	**0.858**	**0.784**	**0.731**	**0.867**	**0.829**	**0.795**

The obtained results indicate that pre-trained models on simulated data exhibit better adaptability to variations in contrast and noise in data obtained from different sites and scanners. This is particularly noticeable in the case of site F and ACDC data-set, which both consist of data acquired from a different scanner vendor compared to the data the model was trained on. In most cases, the features learned from a larger set of variations in simulated data have an impact on the segmentation performance of RV and LVM structures. We attribute this to large-scale variations in RV and LVM structures that are present for both male and female populations in the generated images, not present in the training set. We further explore a minimum size of real MR images required for fine-tuning the pre-trained model to attain acceptable results, but also retain the same results as when trained with a full set of available images. As seen in Table 1, reducing the number of images available for training from scratch significantly impacts the performance of the segmentation models, whereby in some cases, the Dice score is reduced by 9%. Reducing the amount of data available for fine-tuning does impact the segmentation performance, but not as significantly as in models trained from scratch. We perform a similar experiment on the models trained using ACDC data. The results displayed in Table 3 show similar behavior of pre-trained models when fine-tuning for cardiac cavity segmentation on the ACDC data-set. While the improvement in the performance of pre-trained models is not immediately evident for models fine-tuned on larger data-sets, we observe a better generalization performance when a small amount of images is available for fine-tuning compared to models trained from scratch, such as shown in Fig. 4. The same applies to the case of performance retention, which is significantly improved by pre-training.

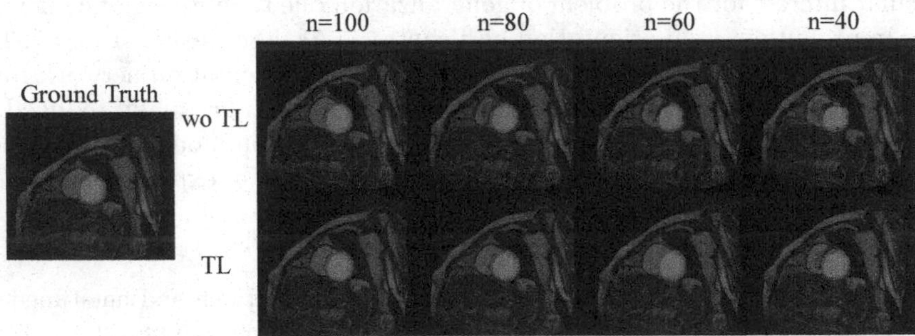

Fig. 4. Real data reduction for comparing the results of networks without (wo TL) and with transfer learning using simulated data and fine tuning with real data.

Table 3. Segmentation performance of the models trained from scratch and fine-tuned on the ACDC dataset. All metrics are average Dice scores obtained for each model.

Train N	Pre-training with simulated data	cCMR Test Set Site A (N=5)			Site B (N=10)			Site C (N=5)			Site D (N=30)			Site E (N=6)			Site F (N=24)			ACDC Test Set N=80		
		LV	RV	LVM	LV	RV	LVM	LV	RV	LVM	LV	RV	LVM	LV	RV	LVM	LV	RV	LVM	LV	RV	LVM
120	NO	0.897	0.872	0.843	0.861	0.833	0.820	0.859	0.862	0.811	0.867	0.848	0.821	0.889	0.852	0.848	0.885	0.872	0.863	**0.941**	**0.922**	**0.907**
	YES	**0.902**	**0.878**	**0.851**	**0.872**	**0.851**	**0.834**	**0.862**	0.863	**0.830**	**0.879**	**0.858**	**0.846**	**00.893**	**0.862**	**0.857**	**0.893**	**0.891**	**0.883**	0.931	0.914	0.897
100	NO	0.893	0.869	0.841	0.859	0.831	0.815	0.855	0.858	0.817	0.857	0.839	0.817	0.881	0.848	0.841	0.864	0.861	0.859	**0.937**	**0.918**	**0.903**
	YES	**0.901**	**0.875**	**0.853**	**0.871**	**0.849**	**0.830**	**0.859**	**0.861**	**0.829**	**0.877**	**0.845**	**0.841**	**0.889**	**0.857**	**0.855**	**0.886**	**0.883**	**0.880**	0.930	0.913	0.895
80	NO	0.881	0.853	0.837	0.852	0.821	0.804	0.843	0.849	0.809	0.848	0.835	0.810	0.876	0.839	0.836	0.841	0.857	0.851	0.918	0.903	0.892
	YES	**0.898**	**0.871**	**0.849**	**0.868**	**0.842**	**0.823**	**0.850**	**0.857**	**0.824**	**0.875**	**0.843**	**0.836**	**0.881**	**0.856**	**0.853**	**0.879**	**0.881**	**0.877**	**0.922**	**0.910**	**0.894**
60	NO	0.875	0.847	0.824	0.843	0.811	0.795	0.828	0.837	0.789	0.835	0.820	0.801	0.859	0.823	0.826	0.827	0.842	0.833	0.897	0.884	0.878
	YES	**0.891**	**0.869**	**0.841**	**0.859**	**0.836**	**0.821**	**0.841**	**0.852**	**0.821**	**0.863**	**0.832**	**0.829**	**0.868**	**0.847**	**0.849**	**0.864**	**0.875**	**0.867**	**0.911**	**0.901**	**0.887**
40	NO	0.859	0.828	0.803	0.831	0.789	0.778	0.808	0.819	0.767	0.820	0.809	0.791	0.836	0.807	0.803	0.812	0.823	0.818	0.875	0.871	0.864
	YES	**0.885**	**0.857**	**0.832**	**0.852**	**0.823**	**0.814**	**0.831**	**0.838**	**0.806**	**0.848**	**0.821**	**0.813**	**0.847**	**0.827**	**0.826**	**0.852**	**0.853**	**0.849**	**0.898**	**0.888**	**0.881**

3 Discussion and Conclusion

In this paper, we investigate a simulation strategy to generate a highly variable, heterogeneous virtual population of CMR images with XCAT-derived subjects characterized by numerous variations in anatomical representation and appearance. Evaluating the usefulness of such a simulated database in a transfer learning setting for the cardiac cavity segmentation task, initial results suggest that simulated data aids with both the generalization capability of the segmentation model to multi-site data, as well as with handling data scarcity. In fact, we show that the segmentation performance on most test cases is retained even when the network is fine-tuned on a smaller number of real data. By reducing the number of real images required for training by almost 80%, the proposed method serves as a promising approach to handle the lack of accurately annotated data in medical imaging, while at the same time increasing the amount of available data for validation and testing.

While these results are promising, an in-depth statistical analysis is needed to establish the effect of this approach in different settings. Furthermore, additional focus should be placed on improving the realism of simulated images, such that the need for fine-tuning is eliminated with the aim to establish a benchmark dataset. In fact, during our experiments, we have conducted a preliminary analysis of the network performance on real MR data without fine-tuning. The results were not satisfactory enough for the network to be used on its own, achieving the highest Dice scores on the ACDC dataset of 0.812, 0.722 and 0.635 for LV, LVM and RV, respectively, motivating us to apply a transfer learning approach. It is also important to note that we have only utilized standard pre-processing techniques on simulated data for pre-training. However, through qualitative analysis of the achieved predictions, we observe that careful design of pre-processing, as well as post-processing approaches, has a potential of improving the segmentation performance of the pre-trained network.

We further hypothesize that additional benefits can be achieved through the investigation on the number of simulated images used in the pre-training stage, as well as the effects of introducing more variation in terms of anatomy and appearance. Not being hampered by data protection and patient privacy, while containing accurate and anatomically meaningful ground-truth, such a heterogeneous data-set can be used as a tool to objectively compare a wide-array of segmentation methods in the literature. A similar approach can be introduced for other anatomies and modalities, especially in settings where data is scarce, as well as for a wider array of medical image processing tasks, besides segmentation. Finally, besides improving the realism of simulated images, we plan to investigate the use of adversarial and domain adaptation methods that could boost the usability of simulated data, even if not completely realistic.

Acknowledgments. This research is a part of the openGTN project, supported by the European Union in the Marie Curie Innovative Training Networks (ITN) fellowship program under project No. 764465.

References

1. Abbasi-Sureshjani, S., Amirrajab, S., Lorenz, C., Weese, J., Pluim, J., Breeuwer, M.: 4D semantic cardiac magnetic resonance image synthesis on XCAT anatomical model. In: Medical Imaging with Deep Learning (2020)
2. Abraham, N., Khan, N.M.: A novel focal Tversky loss function with improved attention U-Net for lesion segmentation. In: 2019 IEEE 16th International Symposium on Biomedical Imaging (ISBI 2019), pp. 683–687. IEEE (2019)
3. Bernard, O., Lalande, A., Zotti, C., Cervenansky, F., et al.: Deep learning techniques for automatic MRI cardiac multi-structures segmentation and diagnosis: is the problem solved? IEEE Trans. Med. Imaging **37**(11), 2514–2525 (2018)
4. Bojorquez, J.Z., Bricq, S., Acquitter, C., Brunotte, F., Walker, P.M., Lalande, A.: What are normal relaxation times of tissues at 3 T? Magn. Reson. Imaging **35**, 69–80 (2017)
5. Chartsias, A., Joyce, T., Dharmakumar, R., Tsaftaris, S.A.: Adversarial image synthesis for unpaired multi-modal cardiac data. In: Tsaftaris, S.A., Gooya, A., Frangi, A.F., Prince, J.L. (eds.) SASHIMI 2017. LNCS, vol. 10557, pp. 3–13. Springer, Cham (2017). https://doi.org/10.1007/978-3-319-68127-6_1
6. Frangi, A.F., Tsaftaris, S.A., Prince, J.L.: Simulation and synthesis in medical imaging. IEEE Trans. Med. Imaging **37**(3), 673 (2018)
7. Goodfellow, I., Pouget-Abadie, J., Mirza, M., Xu, B., et al.: Generative adversarial nets. In: Advances in Neural Information Processing Systems, vol. 27, pp. 2672–2680. Curran Associates, Inc. (2014)
8. He, K., Zhang, X., Ren, S., Sun, J.: Delving deep into rectifiers: surpassing human-level performance on ImageNet classification. In: Proceedings of the IEEE International Conference on Computer Vision, pp. 1026–1034 (2015)
9. Joyce, T., Kozerke, S.: 3D medical image synthesis by factorised representation and deformable model learning. In: Burgos, N., Gooya, A., Svoboda, D. (eds.) SASHIMI 2019. LNCS, vol. 11827, pp. 110–119. Springer, Cham (2019). https://doi.org/10.1007/978-3-030-32778-1_12
10. Razzak, M.I., Naz, S., Zaib, A.: Deep learning for medical image processing: overview, challenges and the future. In: Dey, N., Ashour, A.S., Borra, S. (eds.) Classification in BioApps. LNCVB, vol. 26, pp. 323–350. Springer, Cham (2018). https://doi.org/10.1007/978-3-319-65981-7_12
11. Segars, W., Sturgeon, G., Mendonca, S., Grimes, J., Tsui, B.M.: 4D XCAT phantom for multimodality imaging research. Med. Phys. **37**(9), 4902–4915 (2010)

12. Stanisz, G.J., et al.: T1, T2 relaxation and magnetization transfer in tissue at 3T. Magn. Reson. Med. Official J. Int. Soc. Magn. Reson. Med. **54**(3), 507–512 (2005)
13. Suzuki, K.: Overview of deep learning in medical imaging. Radiol. Phys. Technol. **10**(3), 257–273 (2017). https://doi.org/10.1007/s12194-017-0406-5
14. Wissmann, L., Santelli, C., Segars, W.P., Kozerke, S.: MRXCAT: realistic numerical phantoms for cardiovascular magnetic resonance. J. Cardiovasc. Magn. Reson. **16**(1), 63 (2014)
15. Yi, X., Walia, E., Babyn, P.: Generative adversarial network in medical imaging: a review. Med. Image Anal. **58**, 101552 (2019)

DyeFreeNet: Deep Virtual Contrast CT Synthesis

Jingya Liu[1], Yingli Tian[1(✉)], A. Muhteşem Ağıldere[2], K. Murat Haberal[2], Mehmet Coşkun[2], Cihan Duzgol[3], and Oguz Akin[3]

[1] The City College of New York, New York, NY 10031, USA
ytian@ccny.cuny.edu
[2] Baskent University, 06810 Ankara, Turkey
[3] Memorial Sloan Kettering Cancer Center, New York 10065, USA

Abstract. To highlight structures such as blood vessels and tissues for clinical diagnosis, veins are often infused with contrast agents to obtain contrast-enhanced CT scans. In this paper, the use of a deep learning-based framework, DyeFreeNet, to generate virtual contrast abdominal and pelvic CT images based on the original non-contrast CT images is presented. First, to solve the overfitting issue for a deep learning-based method on small datasets, a pretrained model is obtained through a novel self-supervised feature learning network, whereby the network extracted intensity features from a large-scale, publicly available dataset without the use of annotations and classified four transformed intensity levels. Second, an enhanced high-resolution "primary learning generative adversarial network (GAN)" is then used to learn intensity variations between contrast and non-contrast CT images as well as retain high-resolution representations to yield virtual contrast CT images. Then, to reduce GAN training instability, an "intensity refinement GAN" using a novel cascade intensity refinement strategy is applied to obtain more detailed and accurate intensity variations to yield the final predicted virtual contrast CT images. The generated virtual contrast CTs by the proposed framework directly from non-contrast CTs are quite realistic with the virtual enhancement of the major arterial structures. To the best of our knowledge, this is the first work to synthesize virtual contrast-enhanced abdominal and pelvic CT images from non-contrast CT scans.

Keywords: Virtual contrast CT · Image synthesis · Self-supervised learning · Deep learning

1 Introduction

The use of contrast material is essential for highlighting blood vessels, organs, and other structures on diagnostic tests such as magnetic resonance imaging (MRI) and computed tomography (CT) [7–9]. However, contrast material may cause fatal allergic reactions or nephrotoxicity [1,10]. This paper attempts to seek a dye-free solution by automatically generating virtual contrast-enhanced

N. Burgos et al. (Eds.): SASHIMI 2020, LNCS 12417, pp. 80–89, 2020.
https://doi.org/10.1007/978-3-030-59520-3_9

CTs directly from non-contrast CT images. There are existing studies based on image synthesis to assist the clinic diagnosis [6,12,15,17,19]. Recently, generative adversarial networks (GANs) [16] have shown to be promising for synthesizing medical images; for example, investigators have used GANs to synthesize 3D CT images from 2D X-rays with two parallel encoder-decoder networks [18], to synthesize MR images from non-contrast CT images, and to virtually stain specimens [2]. The synthesis of contrast-enhanced brain MR images from non-contrast or low contrast brain MR images has also been reported [3,5].

In this paper, we focus on developing a new GAN-based framework to synthesize abdominal and pelvic contrast-enhanced CT images from non-contrast CT images, thereby virtually enhancing the arterial structures. Compared to synthesizing contrast-enhanced brain MR images, synthesizing contrast-enhanced CT images for the abdomen and pelvis is more challenging since they contain more feature and intensity variations. CT images are also susceptible to misregistration and there is a lack of multiparametric images to provide additional soft-tissue contrast. Additionally, abdominal and pelvic CT scans usually contain hundreds of CT slices with converging complex organs and soft tissue structures. To predict pixel intensity variations between synthesized contrast and the original non-contrast CTs accurately, the algorithm needs to obtain both local and global features. Lastly, with limited medical imaging data, the algorithm needs to account for overfitting during the training.

The contributions of this paper are summarized in the following three aspects. 1) **Virtual Contrast CT Synthesis.** To the best of our knowledge, this is the first work to synthesize virtual contrast-enhanced CT images from non-contrast abdominal and pelvic CT scans, which is more challenging than synthesizing contrast-enhanced brain MR images [3,5]. 2) **Novel DyeFreeNet Framework.** This framework consists of the self-supervised learning network to obtain a pretrained model followed by high-resolution GANs to extract context features and predict intensity variations based on original non-contract CT images. We used a cascade intensity refinement strategy to train GANs in a progressive manner starting with key texture features and coarse intensity learning, followed by refining the intensity variations. 3) **Self-supervised Learning Pretrained Model.** To avoid overfitting and to allow the model to learn rich representative features, we first employed a novel self-supervised learning network to learn a pretrained model from a large-scale, publicly available dataset without the use of human annotations through classifying four intensity categories.

2 DyeFreeNet

The DyeFreeNet framework is proposed with the following two key aspects in mind. First, the virtual contrast CT image will be a contrast-enhanced version of the original non-contrast CT image whereby critical features of the original non-contrast CT image, such as the texture information of the body, organs, and soft tissues, will be preserved. Second, intensity variations in both local and global features of paired pre/post-contrast CT images will be accounted for in network

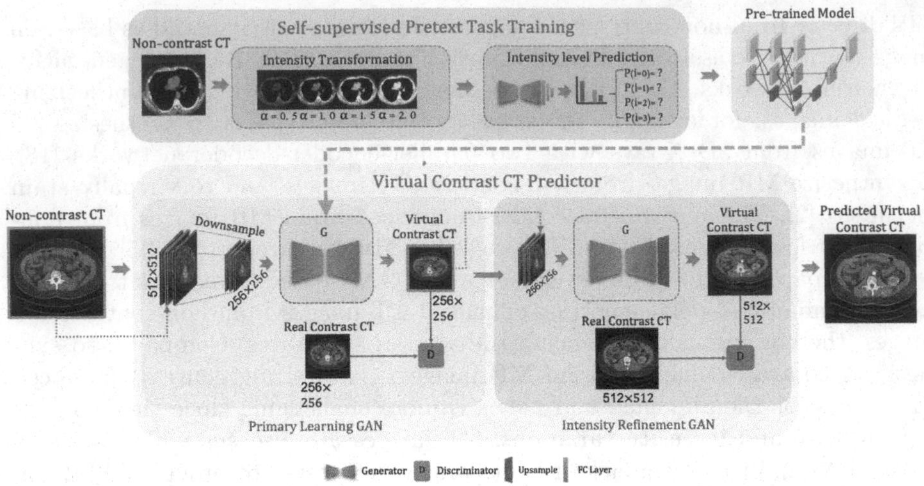

Fig. 1. The DyeFreeNet framework predicts virtual contrast CT images from non-contrast CT images by cascade intensity-learning high-resolution generative adversarial networks (GANs) combining the self-supervised feature learning schema. 1) A self-supervised learning pretrained model is trained by classifying four different intensity levels $(0.5, 1.0, 1.5, 2.0)$ transformed from non-contrast CT images that are available within a large public dataset. 2) A cascade training strategy is employed, whereby the "primary learning generative adversarial network (GAN)" learns the key texture features and coarse intensity variations from non-contrast CT images, and the "intensity refinement GAN" further refines contrast enhancement to yield the final predicted virtual contrast CT images.

design and feature extraction. Figure 1 shows the DyeFreeNet framework that consists of 1) A self-supervised pretrained model for rich feature extraction. 2) High-resolution intensity-learning GANs for preserving high-resolution features and virtual contrast CT generation using a cascade of intensity refinement training strategies.

2.1 Self-supervised Learning Pretrained Model

To speed up the training process and avoid overfitting for virtual contrast CT generation on relatively small dataset, self-supervised learning is proposed to extract rich intensity features from a large-scale, publicly available NLST dataset [11] with non-contrast CT without the use of data annotations and thereby obtain a pretrained model. As shown at Fig. 1, for each non-contrast CT image, intensity variances at four classes $[0, 1, 2, 3]$ are applied by adjusting the intensity coefficient $[0.5, 1.0, 1.5, 2.0], respectively$, to generate transformed CT images. A self-supervised intensity classification network employs the "high-resolution generator" from the virtual contrast CT predictor (see Sect. 2.2 below) as the backbone network for feature extraction. Extracted features are applied for the training of a classifier to predict intensity level, using three fully connected layers with a softmax layer.

The cross-entropy loss function is shown in Eq. 1:

$$loss(c_j|i) = -\frac{1}{K}\sum_{I=0}^{(K-1)} log(F(G(c_j, I)|i)), \tag{1}$$

where the input CT slice c_j is transformed into K levels of intensity I with the coefficient i. F indicates the classification network, and G is the intensity transformation model.

2.2 Virtual Contrast CT Predictor

The virtual contrast CT predictor consisted of cascade intensity refinement to generate high-resolution virtual contrast CT images.

Cascade Training Strategy. Due to the complex structures of abdomen and pelvis CT scans, cascade intensity refinement is split into two stages: 1) The "primary learning GAN" sketched the key texture features and coarse intensity variances. 2) The "intensity refinement GAN" focuses on refining and generating detailed contrast enhancement. The coarse-to-fine procedure utilizes both the spatial and temporal information of CT scans. The "primary learning GAN" takes three consecutive CT slices as input (mimics the RBG channels), and down-samples images to half of the original image size. It then generates the initial contrast CT image, with three continuous CT slices containing rich texture features but insufficient intensity variations serving as the input. The "intensity refinement GAN" takes the initial contrast CT generated by the "primary learning GAN" as the input and obtains more detailed intensity variations. Finally, an up-sampling layer is applied to yield the final predicted high-resolution virtual contrast CT images.

The GAN Architecture. High-resolution features are preserved using a proposed high-resolution encoder-decoder network similar to U-Net [13]. Inspired by High-Resolution Network (HRNet) [14], the encoder learns high-resolution features by back-propagating all layers with current convolutions block L concatenated through all previous convolutional blocks L_{i-1}. Meanwhile, the decoder network was constructed by skip connection, which gradually combines high-level features with low-level features. To preserve rich texture features from the original pre-contrast CT, the feature map from the last block was concatenated with the input features followed by two convolution layers for optimization.

As a "high-resolution generator" extracts the feature representatives, it is essential to map pre-contrast and contrast images in training accurately. We observed that traditional loss functions such as MSE and $L1$ losses easily over-smoothed the generated images. In comparison, GAN solves the issue by applying convolutional networks as discriminator distinguishing the real or generated images. By treating the virtual contrast CT as the result of a gradual approach from the pre-contrast CT to the virtual contrast CT, the virtual contrast

CT can be considered a regression task. The discriminator evaluated MSE and $BCEwithlogits$ losses for the following two sets of feature maps: 1) the feature map extracted from input CT concatenated with the real contrast CT, with the labels as ones; 2) the feature map extracted from input CT concatenated with the virtual CT, with the labels as zeros.

Objective Functions. The "high-resolution generator" learns the mapping between the pre-contrast image x and generated contrast image c to the real contrast image y. The generator G generates virtual contrast CT images, while the discriminator D is trained to distinguish the real and virtual contrast CT image. The objective of the proposed network DyeFreeNet (DFN) is shown as Eq. (2):

$$\mathcal{L}_{DFN}(G, D) = \mathbb{E}_y[LogD(x, y)] + \mathbb{E}_{x,c}[log(1 - D(G(x, c)))], \tag{2}$$

where G aims at minimizing the objective while D maximizes it. An additional MSE loss is appended with the objective to measure the distance between the virtual contrast image with the true image, shown as Eq. (3):

$$\mathcal{L}_{MSE}(G) = \mathbb{E}_{x,y,c}||y - G(x, c)||_2^2. \tag{3}$$

To generate the virtual CT image similar to the real contrast CT image, MSE loss is conducted to optimize the intensity level close to real contrast CT gradually. The loss function of the discriminator is as Eq. (4):

$$\mathcal{L}_{DFN}(D) = D_{MSE}(G(x, c)) + D_{BCE}(G(x, c)). \tag{4}$$

Furthermore, perceptual loss [4] is applied for feature level comparison of generated and real virtual contrast CT images in between all the convolutional blocks. Therefore, the total objective of DyeFreeNet is:

$$G* = arg \min_G \max_D \mathcal{L}_{DFN} + \lambda \mathcal{L}_{MSE}(G) + \sum^{j} \ell_{feat}^{\phi,j}(y', y), \tag{5}$$

where λ adjusts the weight of MSE loss which set as 0.01, and y', y are the virtual and real features from the jth convolutional layer.

3 Experiments

Training and Validation Dataset. We assembled a retrospective CT dataset for examinations. All CT examinations were obtained using a dual-source multi-detector CT scanner. Patients were positioned supine on the table. Pre-contrast imaging of the abdomen was acquired from the dome of the liver to the iliac crest in an inspiratory breath hold by using a detector configuration of 192×0.6 mm, a tube current of 90 kVp, and a quality reference of 277 mAs. After intravenous injection of a 350 mg/ml non-ionic contrast agent (1.5 mL per kg of body weight at a flow rate of 4 mL/s), bolus tracking was started in the abdominal aorta at the level of the celiac trunk with a threshold of 100 HU. Scans were acquired using attenuation-based tube current modulation (CARE Dose 4D, Siemens). We focused on synthesizing the early-stage post-contrast CTs (vascular/arterial phase) from pre-contrast CTs. A total of 4,481 CT slices were used for training and validation while 489 CT slices were used for image quality evaluation.

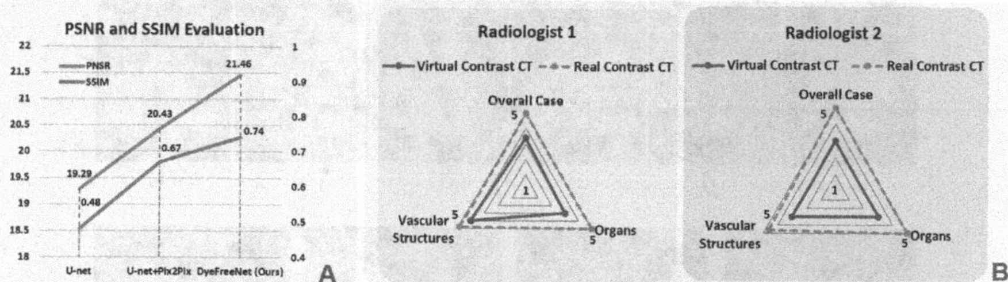

Fig. 2. **A**: The PSNR and SSIM results of the baseline models (U-Net and U-Net+Pix2Pix GAN) and the proposed DyeFreeNet on the comparison between virtual contrast CT images and the real contrast CT images. **B**: The evaluation scores from the assessments of two radiologists in three aspects: overall image quality, image quality of the organs, and the image quality of the vascular structures. Using a 5-point Likert scale from "1" = poor to "5" = excellent, the average evaluation score for virtual contrast CT images is "3" (acceptable). In particular, the image quality of the vascular structures was scored highest as the model was trained with early-stage arterial phase CT images.

Experimental Set Up and Parameter Settings

Self-supervised Learning Pretrained Model. 41589 CT slices of low-dose spiral CT scans were selected from the large public national lung screening trial (NLST) dataset [11] for the training of self-supervised pretrained model. The learning rate is set to $1e^{-6}$ and decreased by 0.1 after 5 epochs updated by Adam optimizer. The total training included 10 epochs with a batch size of 8.

Virtual Contrast CT Predictor. Three consecutive CT slices are fed as inputs to learn essential features in each image as well as between different CT slices. The "primary learning GAN" is initialized using the weights of the self-supervised pretrained model. The learning rate is set to $1e^{-5}$ and decreased by 0.1 after 20 epochs for 25 epochs training with a batch size of 4. The weight decay is $5e^{-4}$ with the Adam optimizer. The speed of virtual contrast image generation is 0.19 s/slice on average with one GeForce GTX 1080 GPU using Python 2.7.

The "intensity refinement GAN" is initialized using the weights of the trained primary learning model. The learning rate is $5e^{-5}$ and decreased by 0.1 after 17 epochs for 20 epochs training with a batch size of 4. The weight decay is $5e^{-4}$ using Adam optimizer. The speed of virtual contrast image generation is 0.24 s/slice on a GeForce GTX 1080 GPU using Python 2.7.

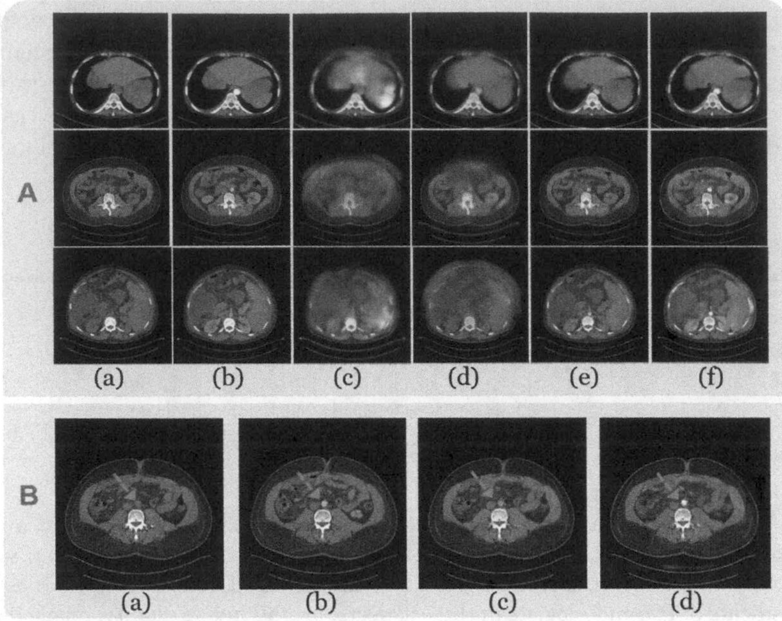

Fig. 3. A: The virtual contrast CT images generated from non-contrast CT images by the proposed DyeFreeNet compared with the baseline models of U-Net and U-Net+Pix2Pix GAN. a) Pre-contrast CT images. b) True contrast CT images as the ground truth. c) Virtual contrast CT images generated by U-Net. d) Virtual contrast CT images synthesized by U-Net+Pix2Pix GAN. e) Virtual contrast CT images synthesized by "primary learning GAN." f) Virtual contrast CT images predicted by "intensity refinement GAN." The illustration shows that our proposed framework could effectively synthesize high-resolution virtual contrast CT images similar to ground truth contrast CT images. **B**: The illustration of the contributions of the self-supervised learning-based pretrained model. (a) Pre-contrast CT image. (b) Real contrast CT image. (c) Virtual contrast CT image without the self-supervised pretrained model. (d) Virtual contrast CT with the self-supervised pretrained model. The red arrow indicates the intensity enhancement of the thoracic aorta region. (Color figure online)

Quantitative Evaluation. Quantitative evaluation was performed between paired synthesized virtual contrast CT images and real contrast CT images that served as the ground truth for baseline models (U-Net and U-Net+Pix2Pix) as well as our proposed DyeFreeNet. Voxel-wise difference and error assessment were conducted using Peak Signal to Noise Ratio (PSNR) while non-local structural similarity was assessed using Structural Similarity Index (SSIM). Figure 2A shows that DyeFreeNet outperformed baseline models of U-Net and U-Net+Pix2Pix on PSNR (by 2.16 and 1.02, respectively), and also on SSIM (by 0.26 and 0.07, respectively).

Qualitative Evaluation by Radiologists. Blind reviews of paired pre-contrast CT images with real contrast CT images and paired pre-contrast CT images and virtual contrast CT images were conducted. Two radiologists independently assessed a total of 489 pairs of real and synthesized images on the following three aspects: overall image quality, image quality of the organs, and image quality of the vascular structures. Qualitative scores were based on a 5-point Likert scale, with scores ranging from "1" = poor, "2" = sub-optimal, "3" = acceptable, "4" = good, and "5" = excellent. Figure 2B shows that the radiologists gave an average score of "3" (acceptable) for overall image quality, image quality of the organs, and image quality of the vascular structures for the virtual contrast images compared with an average score of "5" (excellent) for the real contrast images. The average score for the image quality of the vascular structures was slightly higher than the overall image quality and image quality of the organs. In contrast, the average score was slightly lower for the image quality of the organs, likely because the virtual contrast images were generated from the network trained with vascular (arterial) phase images.

Comparison with Baseline Models. Additional results are illustrated in Fig. 3A. Pre-contrast and real contrast CT images are shown in Fig. 3A(a) and (b), respectively; enhanced regions on the real contrast CT images depict the arterial structures. Virtual contrast CT images generated by U-Net, U-Net+Pix2Pix GAN, "primary learning GAN" of DyeFreeNet, and "intensity refinement GAN" of DyeFreeNet are shown in Fig. 3A(c–f). Although U-Net partially learned the intensity variations, the virtual contrast CT is very blurry and includes artifacts. While traditional MSE or $L1$ loss functions can be applied, they easily oversmooth the predicted image which is problematic as high-resolution images are required for diagnosis. By using the "primary learning GAN," texture features are successfully preserved from the pre-contrast images. However, although the resolution is increased, the intensity variance learning is decreased. Using a pretrained model with "primary learning GAN" results in better feature extraction but the intensity variations still need to be improved. With the "intensity refinement GAN," the DyeFreeNet accurately enhances the vascular structures. The intensity variations and texture features are both learned and preserved in this framework with contributions by the self-supervised learning pre-trained model and the cascade framework.

Comparison with Self-supervised Pre-trained Model. Figure 3B illustrates the results with and without using the self-supervised intensity pretrained model. Pre-contrast and real post-contrast CT images are shown in Fig. 3B(a) and (b). Although the thoracic aorta region (red arrow) is enhanced without using the pretrained model as shown in Fig. 3B(c), with rich feature extraction from the pretrained model, the intensity variations are significantly enhanced as shown in Fig. 3B(d).

Remaining Challenges and Future Work. In this paper, the performance of the model is evaluated at the arterial stage. Our future work will seek to extend the DyeFreeNet network for multi-stage virtual contrast generation (i.e., portal and delayed phases). Its potential limitation is that the misalignment between pre-contrast CT and contrast CT (as the training data) might be more significant than the arterial phase, resulting in generating the artifacts that affect the enhancement accuracy. It is essential to tackle the misalignment issue. Furthermore, to validate the possibility of clinical practice, a downstream task assessment will be developed in future work, such as nodule detection, blood vessel segmentation, and organ segmentation.

4 Conclusion

We developed a self-supervised intensity feature learning-based framework, DyeFreeNet, to automatically generate virtual contrast-enhanced CT images from non-contrast CT images. The rich features extracted by the self-supervised pretrained model and a coarse-to-fine cascade intensity refinement training schema significantly contributed to high-resolution contrast CT image synthesis. The promising results show high potential to generate virtual contrast CTs for clinic diagnosis.

Acknowledgements. This material is based upon work supported by the National Science Foundation under award number IIS-1400802 and Memorial Sloan Kettering Cancer Center Support Grant/Core Grant P30 CA008748.

References

1. Andreucci, M., Solomon, R., Tasanarong, A.: Side effects of radiographic contrast media: pathogenesis, risk factors, and prevention. BioMed Res. Int. **2014**, 741018 (2014)
2. Bayramoglu, N., Kaakinen, M., Eklund, L., Heikkila, J.: Towards virtual H&E staining of hyperspectral lung histology images using conditional generative adversarial networks. In: Proceedings of the IEEE International Conference on Computer Vision, pp. 64–71 (2017)
3. Gong, E., Pauly, J.M., Wintermark, M., Zaharchuk, G.: Deep learning enables reduced gadolinium dose for contrast-enhanced brain MRI. J. Magn. Reson. Imaging **48**(2), 330–340 (2018)
4. Johnson, J., Alahi, A., Fei-Fei, L.: Perceptual losses for real-time style transfer and super-resolution. In: Leibe, B., Matas, J., Sebe, N., Welling, M. (eds.) ECCV 2016. LNCS, vol. 9906, pp. 694–711. Springer, Cham (2016). https://doi.org/10.1007/978-3-319-46475-6_43
5. Kleesiek, J., et al.: Can virtual contrast enhancement in brain MRI replace gadolinium?: a feasibility study. Invest. Radiol. **54**, 653–660 (2019)
6. Li, Z., Wang, Y., Yu, J.: Brain tumor segmentation using an adversarial network. In: Crimi, A., Bakas, S., Kuijf, H., Menze, B., Reyes, M. (eds.) BrainLes 2017. LNCS, vol. 10670, pp. 123–132. Springer, Cham (2018). https://doi.org/10.1007/978-3-319-75238-9_11
7. Litjens, G., et al.: A survey on deep learning in medical image analysis. Med. Image Anal. **42**, 60–88 (2017)

8. Liu, J., Li, M., Wang, J., Wu, F., Liu, T., Pan, Y.: A survey of MRI-based brain tumor segmentation methods. Tsinghua Sci. Technol. **19**(6), 578–595 (2014)
9. Liu, J., Cao, L., Akin, O., Tian, Y.: 3DFPN-HS^2: 3D feature pyramid network based high sensitivity and specificity pulmonary nodule detection. In: Shen, D., et al. (eds.) MICCAI 2019. LNCS, vol. 11769, pp. 513–521. Springer, Cham (2019). https://doi.org/10.1007/978-3-030-32226-7_57
10. Mentzel, H.J., Blume, J., Malich, A., Fitzek, C., Reichenbach, J.R., Kaiser, W.A.: Cortical blindness after contrast-enhanced CT: complication in a patient with diabetes insipidus. Am. J. Neuroradiol. **24**(6), 1114–1116 (2003)
11. National Lung Screening Trial Research Team: The national lung screening trial: overview and study design. Radiology **258**(1), 243–253 (2011). [dataset]
12. Ren, J., Hacihaliloglu, I., Singer, E.A., Foran, D.J., Qi, X.: Adversarial domain adaptation for classification of prostate histopathology whole-slide images. In: Frangi, A.F., Schnabel, J.A., Davatzikos, C., Alberola-López, C., Fichtinger, G. (eds.) MICCAI 2018. LNCS, vol. 11071, pp. 201–209. Springer, Cham (2018). https://doi.org/10.1007/978-3-030-00934-2_23
13. Ronneberger, O., Fischer, P., Brox, T.: U-Net: convolutional networks for biomedical image segmentation. In: Navab, N., Hornegger, J., Wells, W.M., Frangi, A.F. (eds.) MICCAI 2015. LNCS, vol. 9351, pp. 234–241. Springer, Cham (2015). https://doi.org/10.1007/978-3-319-24574-4_28
14. Sun, K., Xiao, B., Liu, D., Wang, J.: Deep high-resolution representation learning for human pose estimation. arXiv preprint arXiv:1902.09212 (2019)
15. Wei, W., et al.: Learning myelin content in multiple sclerosis from multimodal MRI through adversarial training. In: Frangi, A.F., Schnabel, J.A., Davatzikos, C., Alberola-López, C., Fichtinger, G. (eds.) MICCAI 2018. LNCS, vol. 11072, pp. 514–522. Springer, Cham (2018). https://doi.org/10.1007/978-3-030-00931-1_59
16. Yi, X., Walia, E., Babyn, P.: Generative adversarial network in medical imaging: a review. Med. Image Anal. **58**, 101552 (2019)
17. Ying, X., Guo, H., Ma, K., Wu, J., Weng, Z., Zheng, Y.: X2CT-GAN: reconstructing CT from biplanar x-rays with generative adversarial networks. In: Proceedings of the IEEE Conference on Computer Vision and Pattern Recognition, pp. 10619–10628 (2019)
18. Ying, X., Guo, H., Ma, K., Wu, J., Weng, Z., Zheng, Y.: X2CT-GAN: reconstructing CT from biplanar x-rays with generative adversarial networks. In: The IEEE Conference on Computer Vision and Pattern Recognition (CVPR), June 2019
19. Zhao, H.Y., et al.: Synthesis and application of strawberry-like Fe3O4-Au nanoparticles as CT-MR dual-modality contrast agents in accurate detection of the progressive liver disease. Biomaterials **51**, 194–207 (2015)

A Gaussian Process Model Based Generative Framework for Data Augmentation of Multi-modal 3D Image Volumes

Nicolas H. Nbonsou Tegang[1(✉)], Jean-Rassaire Fouefack[1,2,3], Bhushan Borotikar[3,4], Valérie Burdin[2,3], Tania S. Douglas[1], and Tinashe E. M. Mutsvangwa[1,2]

[1] Division of Biomedical Engineering, University of Cape Town, Cape Town, South Africa
{nbnnic001,ffjea001}@myuct.ac.za
{tania.douglas,tinashe.mutsvangwa}@uct.ac.za

[2] Department of Image and Information Processing, IMT-Atlantique, Brest, France
valerie.burdin@imt-atlantique.fr

[3] Laboratory of Medical Information Processing (LaTIM, INSERM U1101), Brest, France
bhushan.borotikar@gmail.com

[4] Symbiosis Centre for Medical Image Analysis, Symbiosis International University, Pune, India

Abstract. Medical imaging protocols routinely employ more than one image modality for diagnosis or treatment. To reduce healthcare costs in such scenarios, research is ongoing on synthetically generating images of one modality from another. Machine learning has shown great potential for such synthesis but performance suffers from scarcity of high quality, co-registered, balanced, and paired multi-modal data required for the training. While methods that do not depend on paired data have been reported, image quality limitations persist including image blurriness. We propose a framework to synthetically generate co-registered and paired 3D volume data using Gaussian process morphable models constructed from a single matching pair of multi-modal 3D image volumes. We demonstrate the application of the framework for matching pairs of CT and MR 3D image volume data with our main contributions being: 1) A generative process for synthesising valid, realistic, and co-registered pairs of CT and MR 3D image volumes, 2) Evaluation of the consistency of the coupling between the generated image volume pairs. Our experiments show that the proposed method is a viable approach to data augmentation that could be used in resource limited environments.

The research has been supported by the Intra-Africa Mobility Scheme of the Education, Audiovisual and Culture Executive Agency at the European Commission and by the South African Research Chairs Initiative of the Department of Science and Innovation and the National Research Foundation of South Africa (Grant No 98788).

N. Burgos et al. (Eds.): SASHIMI 2020, LNCS 12417, pp. 90–100, 2020.
https://doi.org/10.1007/978-3-030-59520-3_10

Keywords: MRI · CT · Matching multi-modal · Gaussian process · Image synthesis

1 Introduction

Multi-modal imaging leverages complementary information in clinical diagnosis and treatment planning. However, the associated costs implications remain a concern in resource-constrained settings. Furthermore, while automation in medical image processing remains a goal and modern machine learning frameworks have shown remarkable progress in this regard, these efforts are limited by shortage of quality labelled data due to patient privacy and ethical considerations. An additional constraint is the labor intensive nature of image labelling. Researchers have turned to medical image synthesis to address the above drawbacks. Medical image synthesis aims to 1) find a representation of an image from a source modality to a target modality, for instance from magnetic resonance imaging (MRI) to computed tomography (CT) (or vice versa) [3,7], or 2) balance the representation of classes in classification problems [12]. In both cases, strategies to resolve the lack of quality image data rely on designing or modifying network architectures to circumvent this bottleneck or on increasing training data with artificially generated images. This latter strategy, termed data augmentation [13,16], can be categorised into different approaches. A common approach capitalizes on traditional image transformations such as rotations, cropping, scaling etc., producing image data that is highly correlated with the original images. Other methods, in similar vein, act on the scale domain (intensity) by adding local or global noise to the image to derive new images [10,13]. Another approach is to exploit model-based morphism through statistical shape modeling (SSM) to sample new "legal" shapes of the organ of interest and for each sampled shape use the original image to propagate intensities to form new images [16]. Finally, a different approach is to make use of the adversarial training of a generative adversarial network (GAN) to generate realistic images [16]. While the latter two approaches have shown superior efficiency in data augmentation, two distinctions persist: 1) SSM approaches enable modeling of the real distribution of the training data; and 2) GAN approaches provide visually realistic looking generated data superior to SSM approaches. The following summary state-of-the-art highlights the advantages and challenges of both these approaches in multi-modal data augmentation.

GANs map a random vector from a simple distribution to a realistic image [5]; a capability exploited for data augmentation notably for cross modality synthesis [6]. The cycleGAN [19], the current state-of-the-art for data augmentation, makes use of unaligned and unpaired image data to generate paired images (in this case MR and CT images) by adding the cycle of consistency on the gradient of an image. However, the use of unpaired data leads to blurriness in the generated synthetic CT (sCT) [8,9]. In sCT generation, cycleGANs often produce unrealistic interfacing of high-contrast regions, for example, the boundaries between bone and air cavities may be replaced by "soft tissue" [2]. This issue is exacerbated by class imbalance, resulting in hallucination of features when the loss function is modelled as distribution matching [4]. A general limitation of GAN and its variants is their propensity to mode collapse which precludes learning the full distribution of the source domain. Additionally, GANs learn distributions with fairly small support compared to the distribution of real data [1]. These problems in GAN-based data generation worsen with the increase of image resolution and also when images are incorrectly aligned or unpaired in the training data set.

On the other hand, morphable SSMs learn the true distribution from training examples such that when sampling from those models, one can generate data similar to the training set. While the literature on the use of SSM for augmenting medical image data is not vast, two examples illustrate common pitfalls to this approach. In the data augmentation context, model-based approaches serve to increase data variability by simulating new shapes and textures of an organ of interest. However, when used on soft tissue, large deformations may not conserve the morphism and combined with the discrete character of SSM may lead to non-plausible samples [16,18]. Most SSMs are discrete models, meaning that they are defined on a fixed domain; this limits their accuracy, making a continuous domain a desirable property.

Here we propose a mechanistic framework for data augmentation from a matching pair (same patient) of multi-modal image volumes (MRI, CT) using Gaussian process morphable models (GPMMs) [11], a state-of-the-art approach for modeling shape and intensity variations. On one hand the proposed approach blends the use of transformations and Gaussian noise at the intensity level in a model-based morphism framework, benefiting from these jointly, or separately. On the other hand, it leverages the continuous properties of GPMMs in the image domain for a pair of registered multi-modal image volumes. Through extensive experiments using quantitative and perception-oriented metrics, we show that our method maintains the integrity of the coupling between the generated images. The kernel parameterization allows us to generate diverse paired data on demand, and in a controllable way. Section 2 provides the formalism of the framework and the process to generate paired image data. Section 3 presents a validation of the framework and an assessment of the quality of generated image data. Section 4 concludes the paper.

2 Method

Let us begin by noting that an image can be seen as a distribution of intensities through a physical or spatial domain $\mathbf{D}$ and all image modalities of the same patient would have the same domain $\mathbf{D}$ under a machine-specific imaging protocol (different machines and different settings) and orientation invariance (i.e. when a patient is imaged in the same position). Thus, removing the machine-specific protocol and spatial orientation permits neighborhood-based voxel correspondence between two image modalities of the same patient, provided the time lapse between the multiple acquisitions is small. Let X and Y

be two volumes from two different modalities of the same patient under spatial orientation invariance. For the same voxel in both modalities, we consider the couple $(x, y) \in (X, Y)$, and $\mathcal{V}(x)$ and $\mathcal{V}(y)$ their respective neighborhood. The neighborhood-based correspondence between X and Y can be defined as

$$\begin{aligned} T_{XY}\colon X &\longrightarrow Y, \\ \mathcal{V}(x) &\longmapsto T_{XY}(\mathcal{V}(x)) = \mathcal{V}(y) \end{aligned}$$

Based on this, we aim to show that the two image modality spaces can be mapped using a deformation field from one to the other, leveraging, T the spatial cross-modality correspondence function.

2.1 Cross Modality Space Mapping

Let the set $\mathcal{S} = \{(X_n, Y_n, T), n = 1 \dots N |\ X_n \subset \Omega, Y_n \subset \Omega'\}$ where X_n and Y_n are paired image volumes of the same patient n in different modalities (Ω for CT and Ω' for MR), and T be a function for establishing the spatial correspondence across image modalities. The cross modality space mapping problem aims to find a set of new paired images $\mathcal{E}$ corresponding to new samples (virtual patients) and is defined as:

$$\mathcal{E} = \{(X_j, Y_j), j = 1, ...m/ \parallel \mathrm{SIM}(X_n, Y_n) - \mathrm{SIM}(X_j, Y_j)\|_2 \leq \epsilon\} \tag{1}$$

where ϵ is a small positive real number and SIM is a similarity measure between image modalities. An image volume X is represented as a couple $(\mathbf{D}_X, \mathbf{I}_X)$, $\mathbf{D}_X$ and $\mathbf{I}_X$ being the image spatial and intensity domain, respectively. We define the cross modality correspondence function T as:

$$\begin{aligned} T\colon \mathbf{D}_X \times \mathbf{I}_X &\longrightarrow \mathbf{D}_Y \times \mathbf{I}_Y, \\ (x, i_x) &\longmapsto T(x, i_x) = (y, i_y). \end{aligned}$$

Where the couple (x, i_x) represents the location and the associated intensity at the voxel level, respectively. Let us define over X, two parametric distributions of a deformation field $(u(\theta), v(\alpha))$ of parameters θ and α, from which new samples $X_{\theta,\alpha} = (u(\theta|\mathbf{D}_X),\ v(\alpha|\mathbf{I}_X))$ can be obtained by random sampling; u is defined on the space domain $u : \Omega \rightarrow \mathbb{R}^3$, and v on the intensity values $v : \mathbb{R} \rightarrow \mathbb{R}$. The other modality is obtained as: $Y_{\theta,\alpha} = T(X_{\theta,\alpha}) = T(u_j(\theta|\mathbf{D}_X), v_j(\alpha|\mathbf{I}_X))$. For a given true (real or original) couple of image modalities (X_o, Y_o), Eq. 1 can be rewritten as follows:

$$\mathcal{E} = \{(X_{\theta,\alpha}, T(X_{\theta,\alpha})) / \parallel \mathrm{SIM}(X_o, Y_o) - \mathrm{SIM}(X_{\theta,\alpha}, T(X_{\theta,\alpha}))\|_2 \leq \epsilon\} \tag{2}$$

The data augmentation problem in this setting can be formalised for a fixed $\epsilon > 0$ as:

$$\begin{cases} \text{find} & (\theta, \alpha) \\ \text{s.t.} & \parallel \mathrm{SIM}(X_o, Y_o) - \mathrm{SIM}((u(\theta|\mathbf{D}_X), v(\alpha|\mathbf{I}_X)), T(u(\theta|\mathbf{D}_X), v(\alpha|\mathbf{I}_X)))\|_2 \leq \epsilon. \end{cases} \tag{3}$$

2.2 Controlled Distribution Within an Image Modality

In the previous section we have illustrated that one image modality can be mapped to another image modality under the existence of the cross modality correspondence function T. In this section, we will show how we can create a controlled distribution within a given modality. For a given image volume X from one modality, a parametric space of possible images can be defined. A given sequence of θ allows us to estimate a parametric mapping from one image volume of the same modality as X, using GPMMs. Furthermore, two parametric distributions can be estimated, one for the spatial domain, and another for the intensity domain; formulated as follows:

$$(u(\theta), v(\alpha)) \sim (\mathcal{G}(\mu_u, k_u), \mathcal{G}(\mu_v, k_v)) \tag{4}$$

where μ_u, μ_v are zero-value functions and k_u, k_v are Gaussian kernel functions: $k(x, y) = s{\cdot}\exp(\|x-y\|^2/\sigma)$,where s is a scale of the deformation, and σ the radius within which the deformations are correlated. $\mathcal{G}(\mu, k)$ forms a Gaussian process (GP). The deformation field $(u(\theta), v(\alpha))$ intrinsically allows us to generate new sample image volumes of the same modality. In the next section we explain how to leverage this transformation for cross-modality synthesis.

2.3 Data Augmentation Process

Given an image volume from the source modality X, we can use a Gaussian kernel to define free-form deformation models (FFDMs) over the intensities and also the domain $\mathbf{D}$ of the source image modality. The deformation fields (u, v) are randomly sampled from those models, and new domain and intensity function instances are obtained, creating a new image volume. Due to the high computational cost of this operation if performed at the original image volume resolution, the source modality image volume can be down-sampled to a lower resolution. The continuous property of the GPMMs over the domain allows the generation of new image volumes from the source modality by applying the sampled deformations from the down-sampled image volume, to a high-resolution image volume from the source modality X. The corresponding image volume in the target modality is obtained through combining the re-scaled and sampled intensity deformation field v (target modality/source modality) with the cross-modality correspondence function to define intensities on the new domain. The workflow is presented in Fig. 1.

3 Experiments and Results

3.1 Data-Set

This study used image volumes of the pelvis region, obtained in two modalities: MR and CT from 5 patients. Image volumes were acquired for radiotherapy treatment from three Hospital sites $H_i, i = 1, 2, 3$ [14]. This data-set is publicly

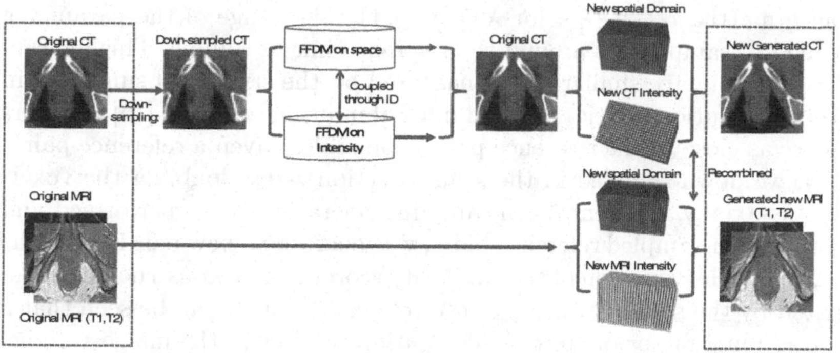

Fig. 1. Workflow to generate new sets of paired image volumes from one original image volume pair. Both deformation models (space and intensity) are built on the lower resolution CT image volume. DF stands for the deformation field sampled from the models. Corresponding DFs from space (DF$_{spatial}$) and intensity (DF$_{int}$) are applied on the high resolution CT image volume to generate new instances of space and intensity. A new CT image volume is obtained by defining the intensities over the new domain. The previous DF$_{int}$ acting on intensities are re-scaled (using original MRI intensity range), and applied over the corresponding original MRI. Thus, each new MRI obtained is in correspondence with a generated CT.

available for research purposes along with the associated details of patient selection, acquisition protocol, computational tools, data format and usage notes [14]. As specified in the source descriptions, T1- and T2-weighted MR image volumes, as well as the CT image volumes, were acquired. The original image sizes were 512×512 in the trans-axial plane for two sites and 384×384 for the third site. All sites H_i used different slice thickness, within and between modalities, except for the third site which used the same slice thickness for MR and CT. Image volumes from both image modalities (CT and MR) were already non-rigidly registered for each patient. In the current study, the pre-processing of image volumes from both modalities consisted of several steps. First, we resized all image volumes to the voxel size $1 \times 1 \times 1$ mm. Second, we cropped each image at the level of the sacral base and at the ischial tuberosity to retain the same anatomical information in every patient. The pre-processing was performed in Amira v6.20 (http://www.fei.com/).

3.2 Impact of Image Resolution on Sampled Deformation Field

To mitigate the computational expense of building a model on a high resolution image volume, we down-sampled the source image volume. In order to understand the impact of different down-sampling factors while maintaining a reasonable deformation field (DF), we built the FFDMs as in Luthi et al. [11], and defined a low rank approximation so that the basis functions were computed within 99% of the total variance for each down-sampled resolution factor. The model parameters α_i were randomly sampled such that they could fit the covariance matrix: $\alpha_i = \frac{2\lambda-1}{4}$ for $\lambda \sim \mathcal{N}(0,1)$. The range of the parameter α_i, was defined to sample DFs around the reference image volume. This guaranteed that the paired image similarity, as measured by the structural similarity index (SSIM) (a perception metric) [17] and mutual information (MI) [15] of generated pairs, were as close to the reference pair as possible. Given a reference pair (MR and CT), we down-sampled in the scan direction z- by doubling the voxel size once. Trans-axially, different down-sampling operations were performed and for each of the down-sampled reference pairs, we generated a new pair from each DF. Figure 2 shows the variation of the similarity scores from reconstructed instances, normalized by the similarity score of the reference pair. It can be seen that from a down-sampling factor of 20, the information present in the image volumes no longer reflects the original image volume. The down-sampling factor should be a smaller value that retains the relevant DF while still reducing the computation cost. Results from Fig. 2 suggest that a down-sampling value less than 10 would provide a DF encapsulating almost the same information as the one at high resolution (reference). For the experiments that followed, a down-sampling factor value of 2 was used, motivated by an acceptable computational time.

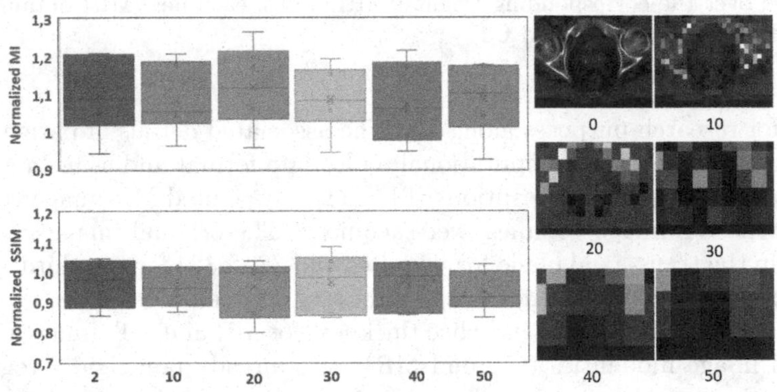

Fig. 2. Left: Variation of the similarity score metrics of reconstructed image volume pair at different down-sampling scale factors. Right: Slices from the reference image volume at those different scales.

3.3 Testing the Ability of the Process to Recover the Reference Image Volume

In order to validate the framework we assessed if reconstruction of the original image pairs from any generated image pair was possible. However, by definition of the GP, the reference pair is obtained by sampling the mean DF, which is a null vector. Instead of sampling the mean DF, we sampled DFs close to the mean DF for both parameter sequences θ, α defined as $\frac{3\lambda-1}{4}$ for $\lambda \sim \mathcal{N}(0,1)$. This avoided sampling of the mean DF which is independent of the resolution. Each sampled

DF allows the generation of one image volume pair. From 5 reference pairs, we generated 100 samples from each pair and computed the similarities between the MR and CT image volumes. The similarity scores of generated image pairs (MR-CT) is reported in Table 1 for different reference images, with the sampled parameters staying the same. However, the resolution was encoded in the basic functions of the GP. The SSIM was used as the metric to quantify the similarity of the generated pairs versus the similarity of the corresponding original pairs from the perceptual point of view. While MI was used to check the consistency of the shared information among generated pair versus in the real pair. Table 1 provides results for this experiment. It can be seen that the constrained DFs reconstructed paired image volumes in close similarity to those generated from reference pairs; the standard deviation of the similarity measures was in the order of 10^{-4}. The results are consistent per site, with patients $101P$ and $108P$ from site H_1, with $206P$ and $211P$ from site H_2, and $302P$ from the third site, H_3 (Fig. 3).

Table 1. Mean and standard deviation of SSIM and MI for both the reference (real image volume) and generated pairs. The standard deviation is of the order of 10^{-4}. R = Real; G = Generated; and $T1$ and $T2$ stand for T1- and T2-weighted MRI, respectively.

Metrics	Domain pair	Patient				
		101P	108P	206P	211P	302P
SSIM	R (CT, T1)	0.02	0.03	0.32	0.3	0.01
	G (CT, T1)	0.01 ± 2.3	0.03 ± 4.2	0.34 ± 545.4	0.28 ± 287.0	0.01 ± 1.8
	R (CT, T2)	0.03	0.04	0.25	0.38	0.02
	G (CT, T2)	0.04 ± 3.8	0.07 ± 34.0	0.29 ± 627.7	0.33 ± 423.3	0.02 ± 3.0
MI	R (CT, T1)	0.04	0.04	0.06	0.07	0.03
	G (CT, T1)	0.05 ± 14	0.05 ± 11	0.07 ± 6.5	0.07 ± 52.9	0.04 ± 9.12
	R (CT, T2)	0.03	0.04	0.06	0.07	0.04
	G (CT, T2)	0.04 ± 1.8	0.07 ± 2.9	0.07 ± 4.9	0.08 ± 4.5	0.04 ± 12

3.4 Experiment on the Consistency of the Coupling Among Generated Samples

Our approach aimed at generating paired data from one original image volume pair. In this section we report on the assessment of the consistency of the coupling between generated MR and CT. This was done by checking if the generated paired image volumes from an unconstrained model (Fig. 3) maintained the same order of similarity values between the modalities as the original pairs. Figure 4 shows the box plot of the similarity scores of 100 paired image volumes generated from real image volume pairs. The score of each generated image was normalized with the score of the original pair. We can observe that the coupling within

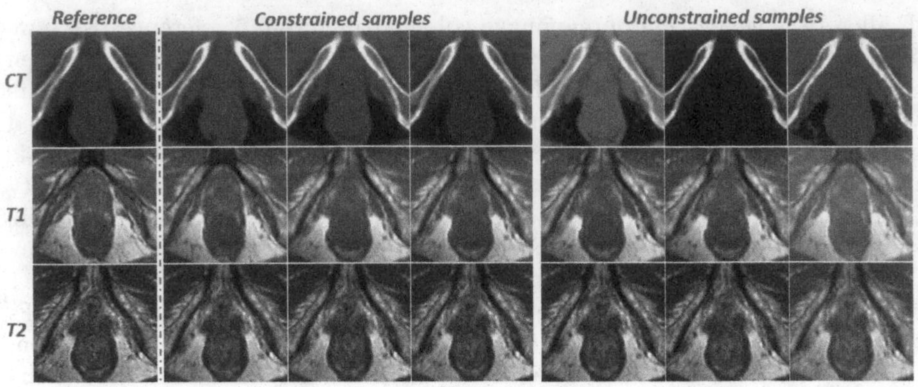

Fig. 3. Samples from the model: Images from the left are obtained from a constrained model, when checking the ability of the model to produce images close to the reference pair. Images from the first column are the real images. The set of images on the right are obtained from random sampling, during an experiment on coupling consistency of generated images. All images have the same shape, however defined on domains of different dimension.

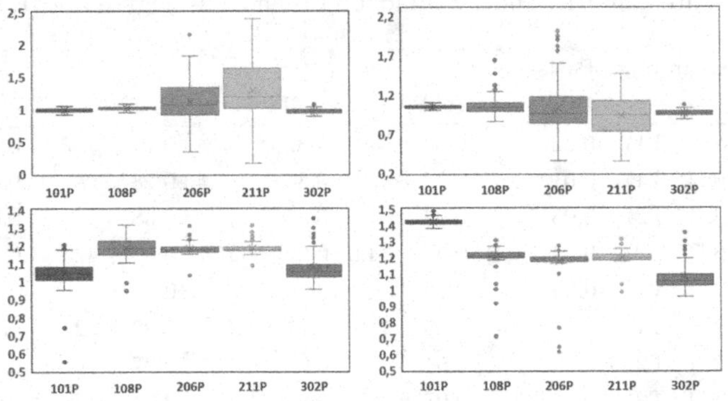

Fig. 4. Normalized similarity scores of 100 sampled image volume pairs (each generated from a real patient pair). Top row: Structural similarity index (CT, T1) left; and (CT, T2) right. Bottom row: adjusted mutual information (CT, T1) left; and (CT, T2) right.

the generated images is consistent because the range of variation of the similarity metrics for samples coming from the same site are similar. Notably, virtual patients, generated from $101P$ and from patient $108P$, both from site H_1, have the same level of similarity scores; the same conclusion can be drawn for patients coming from $206P$ and $211P$, both from H_2. We also observed outliers in the generated image volume pairs indicating that the coupling among the generated image volume was lost or weak. Those outliers represent pairs of images that are far from the reference image, resulting from the sampled transformation mapping the reference CT volume to the a new CT volume. The proposed framework was consistent in keeping the same level of coupling as in the original pairs.

4 Conclusion

In this study, we proposed a generic framework for paired medical image data augmentation, leveraging GPMMs. The performance of the framework was evaluated first on its ability to retrieve the reference pair image volume; second, to generate paired image volumes with similarity scores close to and diverse from the reference volume. The method provides a robust and principled way to increase the diversity of the generated image volume pairs through the Gaussian kernels which can act separately on the spatial domain, or intensity domain, or act together, on the reference image volume. Subsequently, by using different kernels, the diversity of the generated data can be increased. The framework may offer an alternative in cases where paired data is lacking or there is weak diversity in real paired data, permitting valid input for deep learning model training. Future work aims at additional validation of paired synthesized images and assessment of their diversity, by training a deep learning model such as a GAN to ascertain if the approach improves cross-modality synthesis. We also aim to provide a detailed comparison with other state-of-the-art methods using common data-sets.

References

1. Arora, S., Risteski, A., Zhang, Y.: Do GANs learn the distribution? Some theory and empirics. In: International Conference on Learning Representations (2018)
2. Bauer, D.F., et al.: Synthesis of CT images using CycleGANs: enhancement of anatomical accuracy. In: International Conference on Medical Imaging with Deep Learning. Proceedings of Machine Learning Research, vol. 102. PMLR (2019)
3. Burgos, N., et al.: Attenuation correction synthesis for hybrid PET-MR scanners: application to brain studies. IEEE Trans. Med. Imaging **33**(12), 2332–2341 (2014)
4. Cohen, J.P., Luck, M., Honari, S.: Distribution matching losses can hallucinate features in medical image translation. In: Frangi, A.F., Schnabel, J.A., Davatzikos, C., Alberola-López, C., Fichtinger, G. (eds.) MICCAI 2018. LNCS, vol. 11070, pp. 529–536. Springer, Cham (2018). https://doi.org/10.1007/978-3-030-00928-1_60
5. Goodfellow, I., et al.: Generative adversarial nets. In: Advances in Neural Information Processing Systems, pp. 2672–2680 (2014)
6. Han, C., et al.: Combining noise-to-image and image-to-image GANs: Brain MR image augmentation for tumor detection. IEEE Access **7**, 156966–156977 (2019)
7. Han, X.: MR-based synthetic CT generation using a deep convolutional neural network method. Med. Phys. **44**(4), 1408–1419 (2017)
8. Hiasa, Y., et al.: Cross-modality image synthesis from unpaired data using CycleGAN. In: Gooya, A., Goksel, O., Oguz, I., Burgos, N. (eds.) SASHIMI 2018. LNCS, vol. 11037, pp. 31–41. Springer, Cham (2018). https://doi.org/10.1007/978-3-030-00536-8_4
9. Jin, C.B., et al.: Deep CT to MR synthesis using paired and unpaired data. arXiv preprint arXiv:1805.10790 (2018)

10. Lopes, R.G., Yin, D., Poole, B., Gilmer, J., Cubuk, E.D.: Improving robustness without sacrificing accuracy with patch gaussian augmentation. arXiv preprint arXiv:1906.02611 (2019)
11. Lüthi, M., Gerig, T., Jud, C., Vetter, T.: Gaussian process morphable models. IEEE Trans. Pattern Anal. Mach. Intell. **40**(8), 1860–1873 (2017)
12. Momeni, S., et al.: Data augmentation using synthetic lesions improves machine learning detection of microbleeds from MRI. In: Gooya, A., Goksel, O., Oguz, I., Burgos, N. (eds.) SASHIMI 2018. LNCS, vol. 11037, pp. 12–19. Springer, Cham (2018). https://doi.org/10.1007/978-3-030-00536-8_2
13. Nalepa, J., Marcinkiewicz, M., Kawulok, M.: Data augmentation for brain-tumor segmentation: a review. Front. Comput. Neurosci. **13** (2019)
14. Nyholm, T., et al.: MR and CT data with multiobserver delineations of organs in the pelvic area–part of the gold atlas project. Med. Phys. **45**(3), 1295–1300 (2018)
15. Pedregosa, F., et al.: Scikit-learn: machine learning in python. J. Mach. Learn. Res. **12**(85), 2825–2830 (2011). http://jmlr.org/papers/v12/pedregosa11a.html
16. Tang, Z., Chen, K., Pan, M., Wang, M., Song, Z.: An augmentation strategy for medical image processing based on statistical shape model and 3D thin plate spline for deep learning. IEEE Access **7**, 133111–133121 (2019)
17. Wang, Z., Bovik, A.C., Sheikh, H.R., Simoncelli, E.P.: Image quality assessment: from error visibility to structural similarity. IEEE Trans. Image Process. **13**(4), 600–612 (2004)
18. Wu, S., Nakao, M., Tokuno, J., Chen-Yoshikawa, T., Matsuda, T.: Reconstructing 3D lung shape from a single 2D image during the deaeration deformation process using model-based data augmentation. In: 2019 IEEE EMBS International Conference on Biomedical & Health Informatics (BHI), pp. 1–4. IEEE (2019)
19. Zhu, J.Y., Park, T., Isola, P., Efros, A.A.: Unpaired image-to-image translation using cycle-consistent adversarial networks. In: Proceedings of the IEEE International Conference on Computer Vision, pp. 2223–2232 (2017)

Frequency-Selective Learning for CT to MR Synthesis

Zi Lin[1], Manli Zhong[1], Xiangzhu Zeng[2], and Chuyang Ye[1(✉)]

[1] School of Information and Electronics, Beijing Institute of Technology, Beijing, China
chuyang.ye@bit.edu.cn

[2] Department of Radiology, Peking University Third Hospital, Beijing, China

Abstract. *Magnetic resonance* (MR) and *computed tomography* (CT) images are important tools for brain studies, which noninvasively reveal the brain structure. However, the acquisition of MR images could be impractical under conditions where the imaging time is limited, and in many situations only CT images can be acquired. Although CT images provide valuable information about brain tissue, the anatomical structures are usually less distinguishable in CT than in MR images. To address this issue, *convolutional neural networks* (CNNs) have been developed to learn the mapping from CT to MR images, from which brains can be parcellated into anatomical regions for further analysis. However, it is observed that image synthesis based on CNNs tend to lose information about image details, which adversely affects the quality of the synthesized images. In this work, we propose frequency-selective learning for CT to MR image synthesis, where multiheads are used in the deep network for learning the mapping of different frequency components. The different frequency components are added to give the final output of the network. The network is trained by minimizing the weighted sum of the synthesis losses for the whole image and each frequency component. Experiments were performed on brain CT images, where the quality of the synthesized MR images was evaluated. Results show that the proposed method reduces the synthesis errors and improves the accuracy of the segmentation of brain structures based on the synthesized MR images.

Keywords: Deep learning · Cross-modal synthesis · Multi-frequency domain

1 Introduction

Magnetic resonance (MR) and *computed tomography* (CT) images are important tools for brain studies, which noninvasively reveal the brain structure. MR images, such as T1-weighted and T2-weighted images, can provide information about anatomical structures in the brain [9]. CT is fast in examining lesions and brain morphology and conditions, and it has high sensitivity to bleeding.

N. Burgos et al. (Eds.): SASHIMI 2020, LNCS 12417, pp. 101–109, 2020.
https://doi.org/10.1007/978-3-030-59520-3_11

Thus the CT image is widely used clinically [5]. Both of MR and CT images have advantages in clinical diagnosis, and they provide information for clinical diagnosis in different ways. However, the acquisition of MR images could be impractical under conditions where the imaging time is extremely limited, and in many situations CT images can be acquired. To address this issue, *convolutional neural networks* (CNNs) can be used to synthesize MR images from CT images, and from the synthesized MR images brains can be parcellated into anatomical regions for further analysis.

In the initial research, many groups have made great progress by establishing single- and multi-atlas-based approaches and using non-rigid registration methods to synthesize cross-modal medical images. However, with the continuous development of deep learning in recent years, it has achieved remarkable results in building nonlinear mapping models. Previous works have explored synthesis of various modalities of medical images using CNNs. For example, Nie et al. [10] have developed a *fully convolutional neural network* (FCNN) trained with an adversarial strategy to generate CT images from MR images. Wolterink et al. [13] have used unpaired brain MR and CT images to train a CycleGAN [1] model to generate CT images from MR images. Using a similar strategy, Chartsias et al. [4] have achieved excellent performance in the synthesis of cardiac MR images from cardiac CT images. Zhao et al. [15] proposed an improved U-Net to synthesize MR from CT, and obtained good results of gray matter segmentation.

Despite the promising results given by these CNN-based approaches, the CNN models tend to produce smoothed results for image synthesis, where image details can be lost. This is because in the network model learning, the network model tends to use the low-frequency components that are easier to learn [7]. But the high-frequency component, which contains a lot of detailed information, plays an important role in the anatomical segmentation of brain images. In image synthesis, some researches have achieved excellent results by using frequency-domain information to enhance the network learning strategy [3,6,14]. Therefore, in this work we extend the network developed in [15] and propose frequency-selective learning for CT to MR image synthesis, so that details in the synthesized image can be better preserved. Specifically, the MR images are divided into a high frequency component and a low frequency compponent, and multiheads are used in the deep network for learning the mapping of different frequency components. The outputs of these heads are then combined to produce the final synthesized image, and during network training the loss for each frequency component is combined with the loss for the complete image. To validate the proposed method, we performed experiments with paired CT and MR brain images. The experimental results indicate improved quality of the synthesized image and the segmentation of anatomical structures based on the MR image synthesized by the proposed method.

2 Methods

2.1 Frequency-Selective Learning

The key to the synthesis of MR images from CT images lies in how to obtain a mapping from the domain of CT images to the domain of MR images. CNNs have been shown as an effective way of learning such a mapping [15]. Given a CT image x, a CNN model parametrized by ϕ maps x to an MR image y, where the mapping is mathematically expressed as $y = f(x; \phi)$. The CNN model can be trained by minimizing the loss function that characterizes the differences between the network prediction y and the true MR image $\tilde{y}$. For example, the squared ℓ_2-norm of the difference can be used as the loss function:

$$\mathcal{L} = ||y - \tilde{y}||_2^2. \tag{1}$$

Although such CNN-based method [15] has shown promising results for CT to MR image synthesis, the minimization of the loss in Eq. (1), or its variants, could lead to poor preservation of image details [3,6,14]. The loss of details could then adversely affect the segmentation of brain tissue.

To address this problem, we propose to perform frequency-selective learning for CT to MR image synthesis, where multiheads are used in the deep network for learning the mapping of different frequency components. In the network training, we partition the target MR image into two components, a high-frequency image y_{H} and a low-frequency image y_{L}. Specifically, a Gaussian low-pass filter is applied to the training MR images to obtain y_{L}, and the high-frequency component is obtained by subtracting the low-frequency component from the original image: $y_{\mathrm{H}} = y - y_{\mathrm{L}}$. The network predicts these two components and they are added to give the final output of the network, i.e., the final MR image. The losses for the whole image as well as these two components are jointly used for training the network, so that high-frequency is emphasized in the synthesis.

The prediction for multiple frequency components is achieved via a multi-head structure. Specifically, we can make modifications to a backbone synthesis network architecture, for example, the one proposed in [15]. The prediction layers for each frequency components take features generated by shared layers, and the results are added to form the final MR image.

For each frequency component, as well as the final synthesized image, the *mean squared error* (MSE) is used to measure the reconstruction loss. The complete loss function for network training is the weighted sum of these three reconstruction losses:

$$\mathcal{L}_{\mathrm{All}} = \mathcal{L} + a\mathcal{L}_{\mathrm{L}} + b\mathcal{L}_{\mathrm{H}}, \tag{2}$$

where $\mathcal{L}_{\mathrm{All}}$ is the total loss, $\mathcal{L}$ is the loss for the complete generated image, $\mathcal{L}_{\mathrm{H}}$ is the loss for the generated high-frequency component, and $\mathcal{L}_{\mathrm{L}}$ is the loss for the generated low-frequency component. The hyperparameters, a and b, are used to adjust the weights of the loss terms. By adjusting these weights in the loss function, CNNs can effectively pay attention to the details of the image during the learning process, which leads to better MR synthesis.

2.2 Proposed Network Architecture

The detailed structure of the proposed network is shown in Fig. 1 and described here. Our network is based on a variant of the U-net developed in [15] specifically for CT to MR image synthesis, which has proven to be effective for learning the mapping.

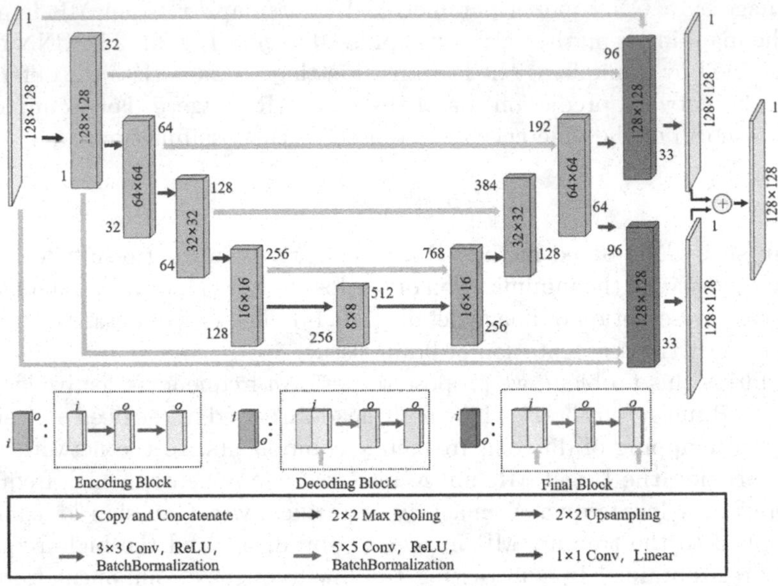

Fig. 1. The network structure for CT to MR image synthesis with frequency-selective learning, where i is the number of input channels for each block and o is the number of output channels of each block.

Like [15], our network is 2D and patch-based. It consists of an encoding part and a decoding part. The encoder part has the same structure as the standard 2D U-net. Each scale contains two 3×3 convolutional layers, which are activated by a *rectified linear unit* (ReLU) [2] and followed by a batch normalization [12] layer. A 2×2 maximum pooling operation is used for downsampling. We used a 2×2 upsampling layer followed by a 2×2 convolution layer and a 3×3 convolution layer in each step of the decoder except for the last step. Each of the two convolutional layers are followed by a batch normalization layer and is activated by ReLU. The last step consists of a multi-head structure, which produces low frequency and high frequency components. Each head contains a 2×2 upsampling layer, and then a 5×5 convolution layer and a 3×3 convolution layer activated by ReLU. For the last 3×3 convolutional layer, considering that the CT value is of great reference significance, like [15] merge the original CT patch before the last convolutional layer. At the end of each head is a 1×1 convolution layer with linear activation. At the end of the entire network structure, we add the output of each head to obtain the final generated image.

2.3 Implementation Details

All MR images are preprocessed before being fed into the network. First, the intensities of MR images are normalized to be in the range of [0,255] using Freesurfer [11]. Then, for each pair of CT and MR data, we rigidly transform the CT image into the MNI space [8], and the MR images are rigidly registered to the corresponding transformed CT images. The axial slices are used for the 2D synthesis. The kernel size for the Gaussian low pass filter that computes the low-frequency components of MR images is set to 25 voxels.

In the training phase, all images are partitioned into 128 × 128 patches with overlapping, where the size of the overlapped region is 128 × 64 voxels. These patches are used to train the proposed network described in the previous section, which generates 128 × 128 MR image patches from 128 × 128 CT image patches. In the test phase, the same patch settings are used, and the generated MR patches are stitched to generate the final MR image by averaging the overlapping part.

The network model was trained using the Adam optimizer, with 30 epoches and 10000 steps per epoch. The batch size is 64. We implemented our method with TensorFlow-based Keras, and model training and testing were performed on an NVIDIA Titan X GPU.

3 Experiments and Results

The experimental data includes 56 pairs of CT and MR images, and each pair corresponds to a different patient. There are 36, 10, and 10 image pairs used for training, validation, and test, respectively. The dimensions of the raw MR and CT images are 512 × 512 × 24 and 364 × 436 × 37, respectively, and the voxel sizes of them are approximately 0.468 mm × 0.468 mm × 5.5 mm and 0.468 mm × 0.468 mm × 5 mm, respectively. The CT and MR images were spatially normalized as described in Sect. 2.3, and all images were resampled to 0.5 mm × 0.5 mm × 5 mm.

Table 1. The list of the structures of interest in the experiments

Left and Right Cerebral White Matter	Left and Right Cerebral Cortex
Left and Right Lateral Ventricle	Left and Right Thalamus Proper
Left and Right Caudate	Left and Right Putamen
Left and Right Pallidum	Left and Right Hippocampus
Left and Right Amygdala	Left and Right Accumbens area
Left and Right VentralDC	CSF

For network training, both CT and MR images were used. For testing the proposed method, only CT images were used as network input, and the MR images were used as the gold standard. For evaluation, we compared the intensities of synthesized and gold standard MR images, where the MSE and *structural similarity* (SSIM) were measured. In addition, because MR images were generally used for brain parcellation, we also compared the brain parcellation results based on the synthesized and gold standard MR images, where the Dice coefficients were computed for each brain region. Specifically, Freesurfer [11] was used for the segmentation, and the structures of interest are listed in Table 1.

The proposed method was compared with the baseline network for CT to MR synthesis developed in [15]. Since the proposed method introduces additional parameters compared with the baseline network, we also considered a broad version of the baseline network, where the penultimate layer is widened so that the total number of network weights are comparable between the proposed network and the broad baseline network. This broad baseline network is referred to as baseline-B for convenience.

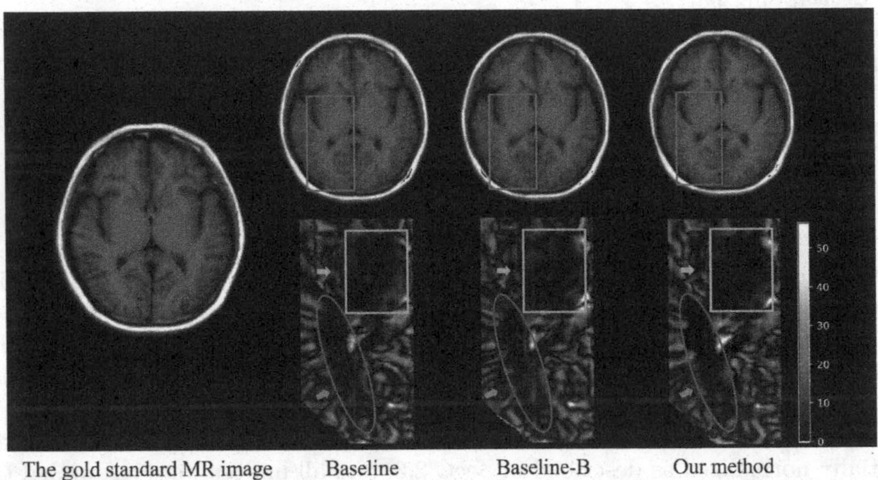

Fig. 2. Examples of synthesized images (top row). The absolute errors in the areas highlighted by red boxes are also shown (bottom row). Note the regions pointed by arrows for comparison. The gold standard is also shown for reference. (Color figure online)

We first determined the hyperparameters a and b in Eq. (1). Specifically, we considered the cases where $a, b \in \{0, 0.01, 0.1, 1, 10\}$ for network training, which led to 25 different combinations of a and b. Then, we computed the mean Dice coefficient for each brain region parcellated from the synethesized MR images of the validation set. To select the most desirable combination, the performance of the proposed method and the baseline and baseline-B methods were compared using the validation set. For each case, we counted the number of structures where the proposed method outperformed both competing methods in terms of average Dice coefficients. The case of $(a, b) = (0.1, 10)$ corresponds to the best number and this combination was used for the rest of the experiments.

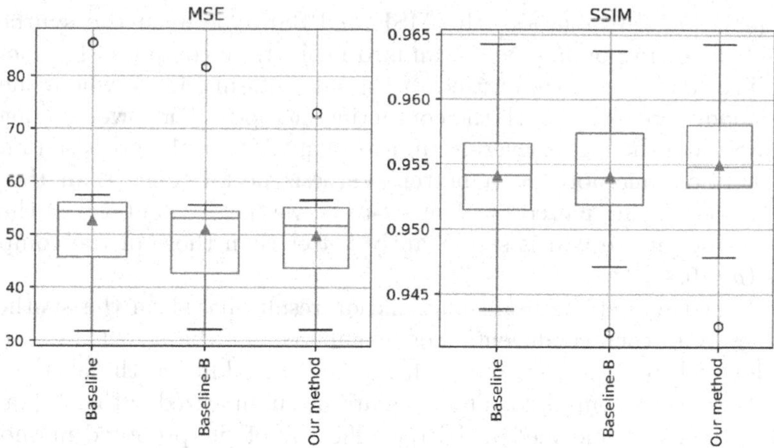

Fig. 3. Boxplots of the MSE and SSIM results of the test subjects. The mean values are also indicated by the green triangles. (Color figure online)

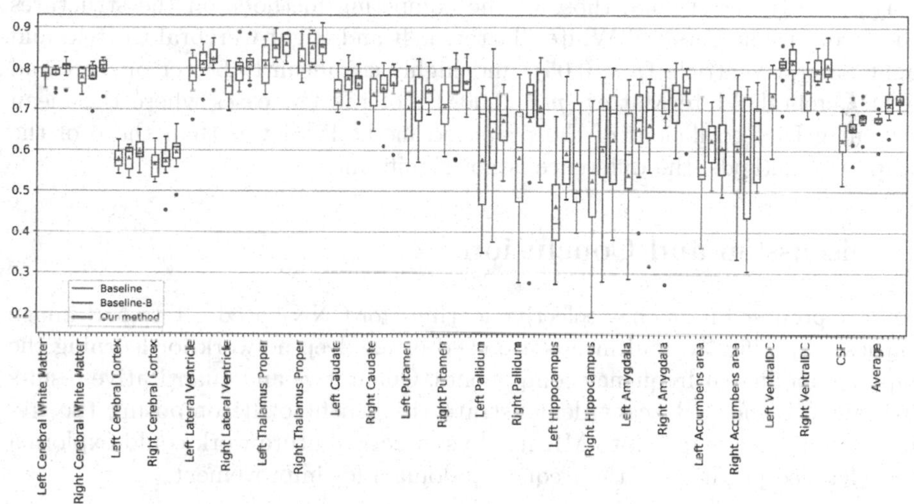

Fig. 4. Boxplots of the Dice coefficients for the test subjects. The mean values are also indicated as the green triangle. (Color figure online)

With the selected hyperparameters, we applied the proposed method to the test data. Examples of the synthesized MR images are shown in Fig. 2, and the results of the gold standard and the competing methods are also shown. For comparison, we computed the mean absolute error between the gold standard and the synthetic image, and the error maps are shown in Fig. 2 for comparison. From the highlighted regions, we can see that the proposed method achieves smaller errors than the competing methods.

For quantitative evaluation, the MSE and SSIM between the synthesized image and the corresponding gold standard image were computed for each test subject. The results are summarized in the boxplots in Fig. 3, where the proposed method is compared with the competing methods. The lower average and median MSE and the higher average and median SSIM of the proposed method indicate that our method has a better synthesis performance than the competing methods. Using paired Student's t-tests, we have also observed that the SSIM result of our method is significantly better than those of the competing methods ($p < 0.01$).

Next, we compared the brain parcellation results based on the synthesized MR images. The Dice coefficients were computed for the structures listed in Table 1 for each test subject and each method. In addition, the average Dice coefficients were also computed. These results are summarized in Fig. 4. For most structures, the mean and median Dice coefficients of the proposed method are higher than those of the competing methods, indicating that our method better synthesizes MR images. In addition, we performed paired Student's t-tests to compare the parcellation results, and the performance of the proposed method is significantly better than those of the competing methods on the structures of Left and Right Cerebral White Matter, Left and Right Cerebral Cortex, and Right Lateral Ventricle ($p < 0.05$), and highly significantly better on Left and Right Cerebral White Matter ($p < 0.001$). For the few cases where the mean or median Dice coefficient of the proposed method is lower than those of the competing methods, the difference is not significant.

4 Discussion and Conclusion

We have proposed frequency-selective learning for CNN-based CT to MR image synthesis. Specifically, multiheads are used in the deep network for learning the mapping of different frequency components. Qualitative and quantitative results on brain CT and MR scans demonstrate the benefit of decomposing the frequency components in CT to MR image synthesis. Future work could explore a more detailed partition of the frequency domain for improvement.

Acknowledgment. This work is supported by Beijing Natural Science Foundation (7192108 & L192058) and Beijing Institute of Technology Research Fund Program for Young Scholars.

References

1. Almahairi, A., Rajeswar, S., Sordoni, A., Bachman, P., Courville, A.: Augmented cyclegan: Learning many-to-many mappings from unpaired data. arXiv preprint arXiv:1802.10151 (2018)
2. Arora, R., Basu, A., Mianjy, P., Mukherjee, A.: Understanding deep neural networks with rectified linear units. arXiv preprint arXiv:1611.01491 (2016)

3. Chakrabarti, A.: A neural approach to blind motion deblurring. In: Leibe, B., Matas, J., Sebe, N., Welling, M. (eds.) ECCV 2016. LNCS, vol. 9907, pp. 221–235. Springer, Cham (2016). https://doi.org/10.1007/978-3-319-46487-9_14
4. Chartsias, A., Joyce, T., Dharmakumar, R., Tsaftaris, S.A.: Adversarial image synthesis for unpaired multi-modal cardiac data. In: Tsaftaris, S.A., Gooya, A., Frangi, A.F., Prince, J.L. (eds.) SASHIMI 2017. LNCS, vol. 10557, pp. 3–13. Springer, Cham (2017). https://doi.org/10.1007/978-3-319-68127-6_1
5. Clements, T.W., et al.: Neurocognitive assessment in patients with a minor traumatic brain injury and an abnormal initial CT scan: can cognitive evaluation assist in identifying patients who require surveillance CT brain imaging? Am. J. Surgery **215**(5), 843–846 (2018)
6. Deng, M., Li, S., Barbastathis, G.: Learning to synthesize: splitting and recombining low and high spatial frequencies for image recovery. arXiv preprint arXiv:1811.07945 (2018)
7. Dziedzic, A., Paparrizos, J., Krishnan, S., Elmore, A., Franklin, M.: Band-limited training and inference for convolutional neural networks. arXiv preprint arXiv:1911.09287 (2019)
8. Grabner, G., Janke, A.L., Budge, M.M., Smith, D., Pruessner, J., Collins, D.L.: Symmetric atlasing and model based segmentation: an application to the hippocampus in older adults. In: Larsen, R., Nielsen, M., Sporring, J. (eds.) MICCAI 2006. LNCS, vol. 4191, pp. 58–66. Springer, Heidelberg (2006). https://doi.org/10.1007/11866763_8
9. Mwangi, B., Matthews, K., Steele, J.D.: Prediction of illness severity in patients with major depression using structural MR brain scans. J. Magnet. Resonance Imaging **35**(1), 64–71 (2012)
10. Nie, D., et al.: Medical image synthesis with deep convolutional adversarial networks. IEEE Trans. Biomed. Eng. **65**(12), 2720–2730 (2018)
11. Reuter, M., Rosas, H.D., Fischl, B.: Highly accurate inverse consistent registration: a robust approach. Neuroimage **53**(4), 1181–1196 (2010)
12. Santurkar, S., Tsipras, D., Ilyas, A., Madry, A.: How does batch normalization help optimization? In: Advances in Neural Information Processing Systems, pp. 2483–2493 (2018)
13. Wolterink, J.M., Leiner, T., Viergever, M.A., Išgum, I.: Generative adversarial networks for noise reduction in low-dose CT. IEEE Trans. Med. Imaging **36**(12), 2536–2545 (2017)
14. Zhang, Y., Yap, P.T., Qu, L., Cheng, J.Z., Shen, D.: Dual-domain convolutional neural networks for improving structural information in 3 T MRI. Magnet. Resonance Imaging **64**, 90–100 (2019)
15. Zhao, C., Carass, A., Lee, J., He, Y., Prince, J.L.: Whole brain segmentation and labeling from CT using synthetic MR images. In: Wang, Q., Shi, Y., Suk, H.-I., Suzuki, K. (eds.) MLMI 2017. LNCS, vol. 10541, pp. 291–298. Springer, Cham (2017). https://doi.org/10.1007/978-3-319-67389-9_34

Uncertainty-Aware Multi-resolution Whole-Body MR to CT Synthesis

Kerstin Kläser[1,2(✉)], Pedro Borges[1,2], Richard Shaw[1,2], Marta Ranzini[1,2], Marc Modat[2], David Atkinson[4], Kris Thielemans[3], Brian Hutton[3], Vicky Goh[2], Gary Cook[2], M. Jorge Cardoso[2], and Sébastien Ourselin[2]

[1] Department Medical Physics and Biomedical Engineering, University College London, London, UK
kerstin.klaser.16@ucl.ac.uk

[2] School of Biomedical Engineering and Imaging Sciences, King's College London, London, UK

[3] Institute of Nuclear Medicine, University College London, London, UK

[4] Centre for Medical Imaging, University College London, London, UK

Abstract. Synthesising computed tomography (CT) images from magnetic resonance images (MRI) plays an important role in the field of medical image analysis, both for quantification and diagnostic purposes. Especially for brain applications, convolutional neural networks (CNNs) have proven to be a valuable tool in this image translation task, achieving state-of-the-art results. Full body image synthesis, however, remains largely uncharted territory, bearing many challenges including a limited field of view and large image size, complex spatial context and anatomical differences between time-elapsing image acquisitions. We propose a novel multi-resolution cascade 3D network for end-to-end full-body MR to CT synthesis. We show that our method outperforms popular CNNs like U-Net in 2D and 3D. We further propose to include uncertainty in our network as a measure of safety and to account for intrinsic noise and misalignment in the data.

Keywords: MR to CT synthesis · Multi-resolution CNN · Uncertainty

1 Introduction

Simultaneous positron emission tomography and magnetic resonance imaging (PET/MRI) is an important tool in both clinical and research applications that allows for a multiparametric evaluation of an individual. It combines the high soft-tissue contrast from MRI with radiotracer uptake distribution information obtained from PET imaging. To accurately reconstruct PET images, it is essential to correct for photon attenuation throughout the patient. A multi-center study on brain images has shown that obtaining tissue attenuation coefficients from synthesised computed tomography (CT) images leads to state-of-the-art results for PET/MRI attenuation correction [12]. In recent years, the field of

N. Burgos et al. (Eds.): SASHIMI 2020, LNCS 12417, pp. 110–119, 2020.
https://doi.org/10.1007/978-3-030-59520-3_12

MR to CT synthesis has shifted towards the use of convolutional neural networks (CNNs) that have proved to be a powerful tool in the MR to CT image translation task, outperforming existing multi-atlas-based methods [11,17]. However, the problem full-body MR to CT synthesis has largely remained untackled. In 2019, Ge et al. [5] attempted to translate full-body MR images to CT images by introducing a multi-view adversarial learning scheme that predicts 2D pseudo CT (pCT) images along three axes (i.e. axial, coronal, sagittal). 3D volumes are obtained for each axis by stacking 2D slices together before an average fusion is performed to obtain one final 3D volume. The synthesis performance is then evaluated on sub-regions of the body (lungs, femur bones, spine etc). They do not, however, provide results on the full volume. We propose a novel learning scheme for uncertainty aware multi-resolution MR to CT synthesis of the full body (MultiRes). Multi-resolution learning has been used for many computer vision tasks such as dynamic scene deblurring [15], optical flow prediction [3] and depth map estimation [4]. In the field of medical imaging, multi-resolution learning is a popular method for image classification [9] and segmentation [8]. These methods learn strong features at multiple levels of scale and abstraction, therefore finding the input/output voxel correspondence based on these features. Due to the large image size of full-body acquisitions and physical GPU memory constraints, high-resolution 3D image synthesis networks can only be trained in a patch-wise manner, thus capturing a limited amount of spatial context. We show that incorporating feature maps learned at multiple resolutions results in significantly better pCT images than using high-resolution images alone. As a means of providing a measure of algorithm safety, and to account for the limited number of training samples, we also model uncertainty [16]. It is important to distinguish between two types of uncertainty: *aleatoric* and *epistemic* uncertainty. Aleatoric uncertainty captures the irreducible variance that exists in the data, whereas epistemic uncertainty accounts for the uncertainty in the model [10]. Aleatoric uncertainty can be further subcategorized into *homoscedastic* and *heteroscedastic*. Homoscedastic uncertainty is constant across all input data, while heteroscedastic uncertainty varies across the input data. It is evident that in our setting the aleatoric uncertainty should be modelled as heteroscedastic, as task performance is expected to vary spatially due to the presence of artefacts, tissue boundaries, small structures etc. By training our network with channel dropout we can stochastically sample from the approximate posterior over the network weights to obtain epistemic uncertainty measures. By explicitly modelling for the intrinsic noise in the data via modifications to our network architecture and loss function we can observe the heteroscedastic uncertainty. The network is encouraged to assign high levels of uncertainty to high error regions, providing a means of understanding what aspects of the data pose the greatest challenges.

2 Methods

The main challenge with whole body data is its size, and the fact that a large field of view is necessary to make accurate predictions. Common networks, such

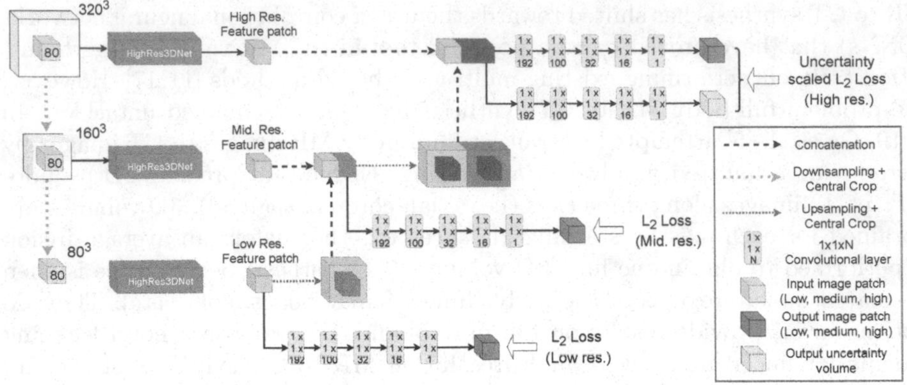

Fig. 1. Proposed MultiRes network architecture. An initial T_1 MR patch of size 320^3 is fed into each instance of the HighRes3DNet architecture [13] at various levels of resolution and field of view. Lower level feature maps are concatenated to those at the next level until the full resolution level, where these concatenated feature maps are passed through two branches consisting of a series of $1 \times 1 \times N$ convolutional layers: one resulting in a synthesised CT patch and the other to the corresponding voxel-wise heteroscedastic uncertainty.

as a U-Net [2], can only store patches of size 160^3 due to GPU memory limitations. This small field of view causes significant issues as it will be demonstrated later in the experiments section. To tackle this issue we propose an end-to-end multi-scale convolutional neural network that takes input patches from full-body MR images at three resolution levels to synthesise high resolution, realistic CT patches. The network also incorporates explicit heteroscedastic uncertainty modelling by casting our task likelihood probabilistically, and epistemic uncertainty estimation via traditional Monte Carlo dropout. We employ a patch-based training approach whereby at each resolution level of the framework a combination of downsampling and cropping operations results in patches of similar size but at different resolutions, spanning varied fields of view. Three independent instances of HighRes3DNet are trained simultaneously, thus not sharing weights, taking patches of each resolution as input each resulting in a feature map with different resolution. Lower level feature maps are concatenated to those at the next level of resolution until the full resolution level, where these concatenated feature maps are passed through two branches of $1 \times 1 \times N$ convolutional layers resulting in a synthesised CT patch and the corresponding voxel-wise heteroscedastic uncertainty. This is illustrated in Fig. 1. We posit, similarly to [8], that such a design allows the network to simultaneously benefit from the fine details afforded by the highest resolution patch and the increased spatial context provided by the higher field of view patches. However, we incorporate an additional level of deep supervision that [8] misses.

2.1 Modelling Heteroscedastic Uncertainty

Previous works on MR to CT synthesis have shown that residual errors are not homogeneously spread throughout the image, rather, they are largely concentrated around organ/tissue boundaries. As such, a heteroscedastic uncertainty model is most suitable for this task, where data-dependent, or intrinsic, uncertainty is assumed to be variable. We begin by modelling our task likelihood as a normal distribution with mean $\boldsymbol{f}^{\boldsymbol{W}}(\boldsymbol{x})$, the model output corresponding to the input $\mathbf{x}$, parameterised by weights $\mathbf{W}$, and voxel-wise standard deviation $\boldsymbol{\sigma}^{\boldsymbol{W}}(\boldsymbol{x})$, the data intrinsic noise:

$$p(\boldsymbol{y}|\boldsymbol{f}^{\boldsymbol{W}}(\boldsymbol{x})) = \mathcal{N}(\boldsymbol{f}^{\boldsymbol{W}}(\boldsymbol{x}), \boldsymbol{\sigma}^{\boldsymbol{W}}(\boldsymbol{x})) \tag{1}$$

Our loss function is derived by calculating the negative log of the likelihood:

$$\begin{aligned} \mathcal{L}(\mathbf{y}, \boldsymbol{x}; \mathbf{W}) &= -\log p(\boldsymbol{y}|\boldsymbol{f}^{\boldsymbol{W}}(\boldsymbol{x})) \\ &\approx \frac{1}{2\boldsymbol{\sigma}^{\boldsymbol{W}}(\boldsymbol{x})^2}\left(\boldsymbol{y} - \boldsymbol{f}^{\boldsymbol{W}}(\boldsymbol{x})\right)^2 + \log \boldsymbol{\sigma}^{\boldsymbol{W}}(\boldsymbol{x}) \\ &= \frac{1}{2\boldsymbol{\sigma}^{\boldsymbol{W}}(\boldsymbol{x})}\mathcal{L}_2(\boldsymbol{y}, \boldsymbol{f}^{\boldsymbol{W}}(\boldsymbol{x})) + \log \boldsymbol{\sigma}^{\boldsymbol{W}}(\boldsymbol{x}) \end{aligned} \tag{2}$$

In those regions where the observed $\mathcal{L}_2$ error remains high, the uncertainty should compensate and also increase. The second term in the loss prevents the collapse to the trivial solution of assigning a large uncertainty everywhere.

2.2 Modelling Epistemic Uncertainty

Test-time dropout has been established as the go-to method for estimating model uncertainty, a Bayesian approximation at inference. By employing dropout during training and testing we can sample from a distribution of sub-nets that in the regime of data scarcity will provide varying predictions. This variability captures the uncertainty present in the network's parameters, allowing for a voxel-wise estimation by quantifying the variance across these samples. In this work, channel dropout was chosen over the traditional neuron dropout. Channel dropout has indeed been shown to be better for convolutional layers where channels fully encode image features while neurons do not encode individually such meaningful information [7]. Dropout samples at inference time are acquired by performing N stochastic forward passes over the network, equivalent to sampling from the posterior over the weights. A measure of uncertainty can be obtained by calculating the variance over these samples on a voxel-wise basis.

2.3 Implementation Details

Experiments were implemented and carried out using NiftyNet, a TensorFlow based deep learning framework tailored for medical imaging [6], and code will be made available on publication. The multi-scale network consists of three residual networks, each taking in an $80 \times 80 \times 80$ MR image patch with different resolutions and fields of view. In order of high, medium, and low resolution, the MR patches are obtained by taking an initial high resolution $320 \times 320 \times 320$ patch and cropping the central $80 \times 80 \times 80$ region (high), downsampling the initial patch by a factor of two and taking the central $80 \times 80 \times 80$ patch (medium), and finally downsampling the initial patch by a factor of four to obtain a $80 \times 80 \times 80$ patch (low).

Starting from the lowest resolution sub-net, the output of size $80 \times 80 \times 80$ is upsampled by a factor of two and centrally cropped. This patch is concatenated with the output of the medium resolution sub-net. This concatenated patch of size $80 \times 80 \times 80 \times 2$ is then upsampled by a factor of two and centrally cropped, before being concatenated to the output of the high-resolution sub-net. These series of upsamplings and crops ensure that the final outputs contain patches with the same field of view prior to the final set of four 3D convolutions of kernel size $1 \times 1 \times 1$, which produces the CT patch.

Heteroscedastic variance is modelled by the addition of a series of four $1 \times 1 \times 1$ convolutional layers following the concatenation of the combined low-medium scale output to the high scale output, architecturally identical to the convolutional layers for the synthesis branch. Channel dropout probability (i.e.: The probability to keep any one channel in a kernel) was set to 0.5, both during training and testing, and $N = 20$ forward passes were carried out for each experiment. The batch size was set to one, ADAM was used as the optimiser and networks were trained until convergence, where this was defined as a sub 5% loss change over a period of 5000 iterations.

3 Experiments and Results

3.1 Data

The dataset used for training and cross-validation consisted of 32 pairs of whole-body MR (voxel size $0.67 \times 0.67 \times 5\,\text{mm}^3$) and CT images (voxel size $1.37 \times 1.37 \times 3.27\,\text{mm}^3$). Whole-body MR images were acquired in four stages. MR pre-processing included bias field correction followed by fusion between stages using a percentile-based intensity harmonisation. All images were resampled to CT resolution. MR and CT images were aligned using first a rigid registration algorithm followed by a very-low-degree-of-freedom non-rigid deformation. A second non-linear registration was performed, using a cubic B-spline with normalised mutual information to correct for soft tissue shift [1,14]. Both CT and MR images were rescaled to be between 0 and 1 for increased training stability.

3.2 Experiments

In addition to our proposed method, we compare results quantitatively and qualitatively against two baselines: U-Net trained with 2D, $224 \times 224 \times 1$,

Table 1. MAE and MSE across all experiments including number of trainable variables. Bolded entries denote best model (p-value < 0.05).

Experiments	Model parameters	MAE (HU)	MSE (HU^2)
2D U-Net (No Unc)	4.84M	112.94 ± 16.04	32081.18 ± 5667.11
3D U-Net (No Unc)	14.49M	99.87 ± 14.17	23217.57 ± 3515.50
MultiRes	2.54M	**62.42 ± 6.8**	**11347.16 ± 3089.12**
MultiRes_{unc}	2.61M	80.14 ± 15.81	14113.83 ± 3668.79

patches with batch size one, and U-Net trained with 3D, 160 × 160 × 160, patches with batch size one. An additional four convolutional layers with kernel size three were added prior to the final 1 × 1 × 1 convolutional layer in the standard architecture as this was found to increase stability during training. All models were trained on the same 22 images while the remaining 10 images were equally split into validation and testing data (Table 1).

3.3 Results

Quantitative Evaluation. The quantitative evaluation consists of a Mean Squared Error ($MSE = \frac{\sum(pCT-CT)^2}{V}$, with V being the total number of non-zero voxels) and Mean Absolute Error ($MAE = \frac{\sum|pCT-CT|}{V}$) analysis between the network outputs and the ground truth CT. We observe that the proposed method without uncertainty performs the best, exhibiting the lowest MAE and MSE averaged across all inference subjects. A paired t-test was performed to show that the results are significantly better (p-value < 0.05). Furthermore, the proposed MultiRes networks show a better performance while decreasing the model size making the networks much more efficient than the U-Net models.

Qualitative Evaluation. Figure 2 shows the ground truth CT and the pCT predictions generated with 2D U-Net, 3D U-Net, proposed MultiRes and proposed MultiRes_{unc} with uncertainty and the subject's MR image as well as the models' corresponding residuals. 2D U-Net clearly cannot capture bone; likely because it lacks the spatial context necessary to construct small (relatively) cohesive structures such as vertebrae. The lungs, having a significantly larger cross-sectional area, are visible, but lack internal consistency. 3D U-Net's bone synthesis is more faithful than its 2D counterpart but is characterised by a large degree of blurriness most evident in the femurs. The proposed MultiRes model exhibits the greatest bone fidelity; the individual vertebrae are more clear, with intensities more in line with what would be expected for such tissues, and the femurs boast more well-defined borders. The proposed MultiRes_{unc} model leads to similar results than the simpler proposed MultiRes model without uncertainty. However, bones are slightly blurrier, likely due to the inclusion of uncertainty term and limited network capacity, but still demonstrates superior bone reconstruction than both U-Net models.

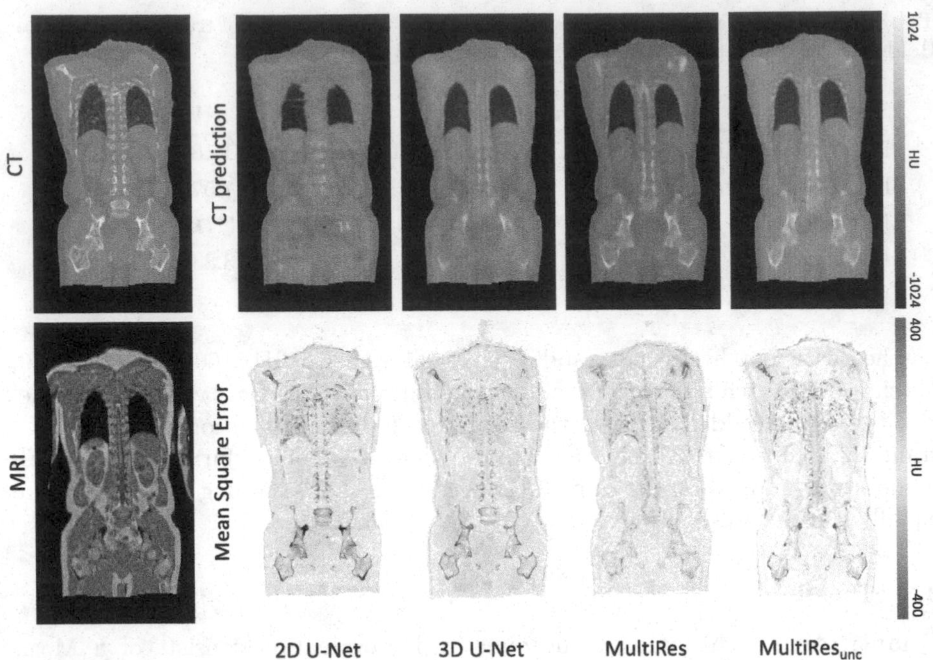

Fig. 2. Top: CT and pCT prediction of 2D U-Net, 3D U-Net, proposed MultiRes and proposed uncertainty aware MultiRes$_{unc}$. Bottom row: MR image and model residuals.

The benefits afforded to MultiRes$_{unc}$ for being uncertainty-aware are showcased in Fig. 3. The joint histograms (a) and (b) are constructed by calculating the error rate, taken as the difference between the ground truth CT and pCT averaged across N=20 dropout samples, at different levels of both epistemic and heteroscedastic uncertainty (standard deviations per voxel) and taking the base 10 log. The red line describes the average error rate at each level of uncertainty. We observe a significant correlation between uncertainty and error rate, suggesting that the model appropriately assigns a higher uncertainty to those regions that are difficult to predict. This correlation is likewise observed when comparing the maps of uncertainty (epistemic: (c), heteroscedastic: (e)) with the corresponding absolute error map (d). Both epistemic and heteroscedastic uncertainties exhibit large values around structure borders, as expected. The borders between tissues are not sharp and there is, therefore, some ambiguity in these regions, which is mirrored by the corresponding overlapping error in the residuals. An increased amount of data should diminish the epistemic uncertainty by providing the network with a greater number of samples from which to learn the correspondence between MR and CT in these areas. The aforementioned blurriness, however, could result in some inconsistency in the synthesis process, which would be captured by the heteroscedastic uncertainty.

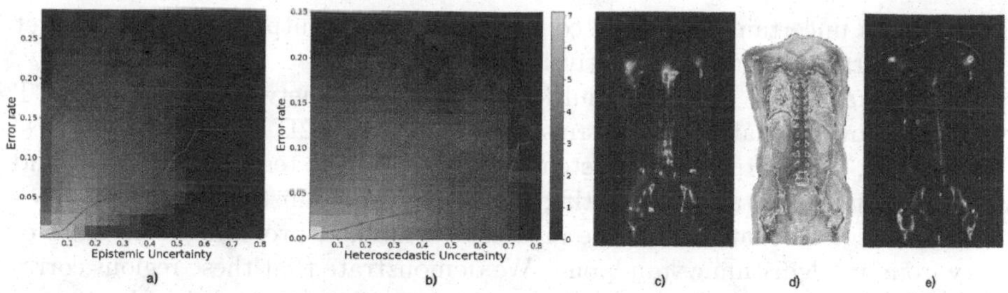

Fig. 3. Joint histogram of prediction uncertainty and error rate for proposed MultiRes$_{unc}$ network: a) Epistemic b) Heteroscedastic. The average error rate at different uncertainty levels is shown by the red line. Error rate tends to increase with increasing uncertainty, showing that the network correlates uncertainty to regions of error. c) Epistemic uncertainty and e) heteroscedastic uncertainty correlate with d) the MAE of the prediction error [0HU, 800HU], solidifying this point. (Color figure online)

Of note is the high degree of uncertainty we observed in the vicinity of air pockets. Unlike corporeal structures, it is expected that these pockets are subject to deformation between the MR and CT scanning sessions, resulting in a lack of correspondence between the acquisitions in these regions. This results in the network attempting to synthesise a morphologically different pocket to what is observed in the MR, resulting in a high degree of uncertainty.

4 Discussion and Conclusions

Our contributions in this work are two-fold: MultiRes, a novel learning scheme for uncertainty aware multi-resolution MR to CT synthesis of the full body, and MultiRes$_{unc}$, a version of this model that incorporates uncertainty as a safety measure and to account for intrinsic data noise. We demonstrate the significantly superior performance (p-value < 0.05) of MultiRes and MultiRes$_{unc}$ by comparing it to single-resolution CNNs, 2D and 3D U-Net, and the importance of modelling uncertainty, showing that MultiRes$_{unc}$ is able to identify regions where the MR to CT translation is most difficult.

In a data-scarce environment, it becomes especially important to quantify uncertainty as networks are unlikely to have sufficient evidence for full convergence.

After all, accurately aligning CT and MR images is inevitable to validate the voxel-wise performance of any image synthesis algorithm until other appropriate methods have been developed that allow validating on non-registered data.

Despite the slightly decreased performance of MultiRes$_{unc}$ compared to MultiRes, both from a quantitative and qualitative standpoint, we posit that the additional insight introduced by modelling uncertainty can compensate for this. Furthermore, while the model does not reconstruct bone-based structures as well as its uncertainty agnostic counterpart, it still outperforms both U-Net models qualitatively and quantitatively.

To summarise, we design a multi-scale/resolution network for MR to CT synthesis, showing that it outperforms single-resolution 2D and 3D alternatives. Furthermore, by incorporating epistemic uncertainty via test time dropout, and heteroscedastic uncertainty by casting the model probabilistically, we can showcase those regions that exhibit the greatest variability, providing a measure of safety from an algorithmic standpoint. We demonstrate that these regions correlate well with the residuals obtained by comparing the outputs with the ground truth, lending further credence to the usefulness of uncertainty's inclusion. We argue that the slight decrease in performance of the uncertainty aware model is insignificant compared to the important additional information provided by the uncertainty.

References

1. Burgos, N., et al.: Attenuation correction synthesis for hybrid PET-MR scanners: application to brain studies. IEEE Trans. Med. Imaging **33**(12), 2332–2341 (2014)
2. Çiçek, Ö., Abdulkadir, A., Lienkamp, S.S., Brox, T., Ronneberger, O.: 3D U-Net: learning dense volumetric segmentation from sparse annotation. In: Ourselin, S., Joskowicz, L., Sabuncu, M.R., Unal, G., Wells, W. (eds.) MICCAI 2016. LNCS, vol. 9901, pp. 424–432. Springer, Cham (2016). https://doi.org/10.1007/978-3-319-46723-8_49
3. Dosovitskiy, A., et al.: FlowNet: learning optical flow with convolutional networks. In: Proceedings of the IEEE International Conference on Computer Vision, pp. 2758–2766 (2015)
4. Eigen, D., Puhrsch, C., Fergus, R.: Depth map prediction from a single image using a multi-scale deep network. In: Advances in Neural Information Processing Systems, pp. 2366–2374 (2014)
5. Ge, Y., Xue, Z., Cao, T., Liao, S.: Unpaired whole-body MR to CT synthesis with correlation coefficient constrained adversarial learning. In: Medical Imaging 2019: Image Processing, vol. 10949, p. 1094905. International Society for Optics and Photonics (2019)
6. Gibson, E., et al.: NiftyNet: a deep-learning platform for medical imaging. Comput. Methods Programs Biomed. **158**, 113–122 (2018)
7. Hou, S., Wang, Z.: Weighted channel dropout for regularization of deep convolutional neural network. In: Proceedings of the AAAI Conference on Artificial Intelligence, vol. 33, pp. 8425–8432 (2019)
8. Kamnitsas, K., et al.: Efficient multi-scale 3D CNN with fully connected CRF for accurate brain lesion segmentation. Med. Image Anal. **36**, 61–78 (2017)
9. Kawahara, J., Hamarneh, G.: Multi-resolution-tract CNN with hybrid pretrained and skin-lesion trained layers. In: Wang, L., Adeli, E., Wang, Q., Shi, Y., Suk, H.-I. (eds.) MLMI 2016. LNCS, vol. 10019, pp. 164–171. Springer, Cham (2016). https://doi.org/10.1007/978-3-319-47157-0_20
10. Kendall, A., Gal, Y.: What uncertainties do we need in Bayesian deep learning for computer vision? In: Advances in Neural Information Processing Systems, pp. 5574–5584 (2017)

11. Kläser, K., et al.: Deep boosted regression for MR to CT synthesis. In: Gooya, A., Goksel, O., Oguz, I., Burgos, N. (eds.) SASHIMI 2018. LNCS, vol. 11037, pp. 61–70. Springer, Cham (2018). https://doi.org/10.1007/978-3-030-00536-8_7
12. Ladefoged, C.N., et al.: A multi-centre evaluation of eleven clinically feasible brain PET/MRI attenuation correction techniques using a large cohort of patients. Neuroimage **147**, 346–359 (2017)
13. Li, W., Wang, G., Fidon, L., Ourselin, S., Cardoso, M.J., Vercauteren, T.: On the compactness, efficiency, and representation of 3D convolutional networks: brain parcellation as a pretext task. In: Niethammer, M., et al. (eds.) IPMI 2017. LNCS, vol. 10265, pp. 348–360. Springer, Cham (2017). https://doi.org/10.1007/978-3-319-59050-9_28
14. Modat, M., et al.: Fast free-form deformation using graphics processing units. Comput. Methods Programs Biomed. **98**(3), 278–284 (2010)
15. Nah, S., Hyun Kim, T., Mu Lee, K.: Deep multi-scale convolutional neural network for dynamic scene deblurring. In: Proceedings of the IEEE Conference on Computer Vision and Pattern Recognition, pp. 3883–3891 (2017)
16. Reinhold, J.C., et al.: Validating uncertainty in medical image translation. In: 2020 IEEE 17th International Symposium on Biomedical Imaging (ISBI), pp. 95–98. IEEE (2020)
17. Wolterink, J.M., Dinkla, A.M., Savenije, M.H.F., Seevinck, P.R., van den Berg, C.A.T., Išgum, I.: Deep MR to CT synthesis using unpaired data. In: Tsaftaris, S.A., Gooya, A., Frangi, A.F., Prince, J.L. (eds.) SASHIMI 2017. LNCS, vol. 10557, pp. 14–23. Springer, Cham (2017). https://doi.org/10.1007/978-3-319-68127-6_2

UltraGAN: Ultrasound Enhancement Through Adversarial Generation

Maria Escobar[1(✉)], Angela Castillo[1], Andrés Romero[2], and Pablo Arbeláez[1]

[1] Center for Research and Formation in Artificial Intelligence, Universidad de los Andes, Bogotá, Colombia
{mc.escobar11,a.castillo13,pa.arbelaez}@uniandes.edu.co
[2] Computer Vision Lab, ETHZ, Zürich, Switzerland
roandres@ethz.ch

Abstract. Ultrasound images are used for a wide variety of medical purposes because of their capacity to study moving structures in real time. However, the quality of ultrasound images is significantly affected by external factors limiting interpretability. We present UltraGAN, a novel method for ultrasound enhancement that transfers quality details while preserving structural information. UltraGAN incorporates frequency loss functions and an anatomical coherence constraint to perform quality enhancement. We show improvement in image quality without sacrificing anatomical consistency. We validate UltraGAN on a publicly available dataset for echocardiography segmentation and demonstrate that our quality-enhanced images are able to improve downstream tasks. To ensure reproducibility we provide our source code and training models.

Keywords: Generative Adversarial Networks · Echocardiography · Ultrasound images · Image quality enhancement

1 Introduction

Echocardiography is one of the most commonly used tests for the diagnosis of cardiovascular diseases as it is low-cost and non-invasive [13]. The quality of echocardiogram images is highly related to three main factors: (i) intrinsic characteristics of the sonograph, (ii) sonograph manipulation [5], and (iii) the manual configurations of the operator. As the intrinsic configurations of each device are hard to control in practice, factors such as the expertise of the user and the contrast adjustment on the machine [23] may potentially affect the final diagnosis. Fundamentally, the criteria to assess the presence of some pathologies or the correct functionality of the left ventricle depends entirely on the quality of the resulting images on this test [11]. Therefore, the interpretability of ultrasound images is limited by their quality [4].

M. Escobar and A. Castillo—Both authors contributed equally to this work.

N. Burgos et al. (Eds.): SASHIMI 2020, LNCS 12417, pp. 120–130, 2020.
https://doi.org/10.1007/978-3-030-59520-3_13

Once physicians acquire echocardiograms, they perform manual segmentation [21] to determine diagnostic measurements of the heart's chambers [13]. The quality of the resulting measurements depends directly on the precise outline of the chambers' boundaries [21]. Nevertheless, if the ultrasound image has a low quality, low resolution, or low contrast, the segmentation will be harder because of the lack of a clear difference between two adjacent structures. With an image that is not entirely intelligible, manual segmentation is purely done under the visual subjectivity of the physician to recognize the heart's boundaries.

Generative Adversarial Networks (GANs) [12] are a type of generative models that learn the statistical representation of the training data. During GAN training, the generator network alternates with the discriminator, so the generator can produce new data that resembles the original dataset. GANs have successfully tackled image-to-image transformation problems, including but not limited to: image colorization [8], super resolution [28], multi-domain and multimodal mappings [26], and image enhancement [7]. In the medical field, several works have introduced GANs into their approach for tasks that include data augmentation [3], medical images attacks [6] and image synthesis [1,2,29].

Recently, some automated methods have been developed for ultrasound image quality enhancement [9,15,16,25]. Liao *et al.* [20] proposed a method to enhance ultrasound images using a quality transfer network. The algorithm was tested in echo view classification, showing that quality transfer helps improve the performance. Also, Lartaud *et al.* [18] trained a convolutional network for data augmentation by changing the quality of the images to create contrast and non-contrast images for segmentation. The augmented data allowed the improvement of the segmentation method. Jafari *et al.* [17] trained a model to transform quality between ultrasound images. The approach introduced a segmentation network in the training of the GAN to provide an anatomical constraint added by the segmentation task. Nevertheless, these methods were developed and evaluated on private data, complicating the possibility of a direct comparison.

In this work, we present UltraGAN, a novel framework for ultrasound image enhancement through adversarial training. Our method receives as input a low-quality image and performs high quality enhancement without compromising the underlying anatomical structures of the input. Our main contributions can be summarized in the two following points: (1) We introduce specific frequency loss functions to maintain both coarse and fine-grained details of the original image. (2) We guide the generation with the anatomical coherence constraint by adding the segmentation map as input in the discriminator.

As our results demonstrate, UltraGAN transfers detailed information without modifying the important anatomical structures. Hence, our method could be useful for a better interpretability in clinical settings. Moreover, for the quantitative validation of our method, we compare the resulting segmentations with and without our enhancement, and we report an increase in performance in this downstream task by adding our enhanced images to the training data. We make our source code and results publicly available[1].

[1] https://github.com/BCV-Uniandes/UltraGAN.

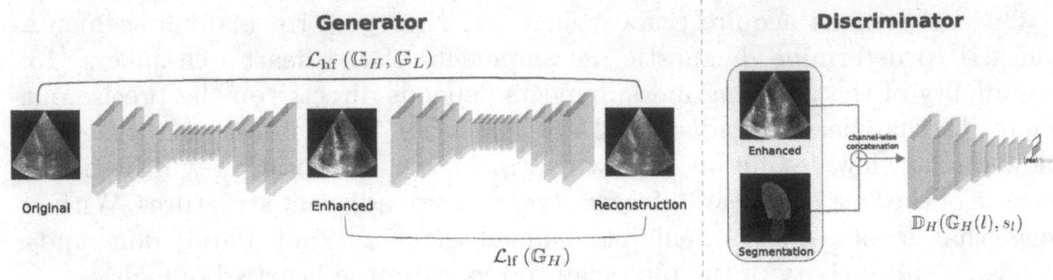

Fig. 1. Overview of our generation scheme. We add a frequency consistency loss to preserve fine details and coarse structures. We concatenate the segmentation map along with the input image for the discriminator to classify as real or enhanced. This particular case corresponds to an enhanced input.

2 Methodology

Our method consists of a Generative Adversarial Network designed to enhance the quality of ultrasounds without compromising the underlying anatomical information contained in the original image. The power of GANs relies on a minimax two-player game, where two different networks are trained in an adversarial fashion. The generator ($\mathbb{G}$) translates images from one domain to another, while the discriminator ($\mathbb{D}$) is in charge of determining whether the image is a real example from the dataset or a generated example coming from $\mathbb{G}$.

2.1 Problem Formulation

Given a set of low-quality ultrasounds $\{l_i\}_{i=1}^N \in L$ with a data distribution $l \sim p_{\text{data}}(l)$ and a set of high quality ultrasounds $\{h_i\}_{i=1}^N \in H$ with a data distribution $h \sim p_{\text{data}}(h)$, our goal is to learn mapping functions that translate from low to high quality domain and vice versa. Accordingly, we have two generators $\mathbb{G}_H : L \rightarrow H$ and $\mathbb{G}_L : H \rightarrow L$. We also have two discriminators: $\mathbb{D}_H$ distinguishes between real high quality images h_i and generated high quality images $\mathbb{G}_H(l_i)$, and $\mathbb{D}_L$ distinguishes between real low-quality images l_i and generated low-quality images $\mathbb{G}_L(h_i)$. Since we want our mapping functions to be aware of the original structural information, we also have the segmentation of the anatomical regions of interest present in the ultrasound image s_h or s_l.

It is important to note that at inference time, we only use the generator trained for the translation from low to high quality, even though we optimized for both generators in the training phase.

2.2 Model

We start from the generator architecture of CycleGAN [30] which consists of a series of down-sampling layers, residual blocks and up-sampling layers. Figure 1 shows the training scheme for our model. For the discriminator, we build upon

PatchGAN [14,30]. Since we want our model to learn how to produce anatomically coherent results, our discriminator has two inputs: the ultrasound image (whether real or generated) and the corresponding segmentation of the anatomical regions of interest.

2.3 Loss Functions

To enforce the task of quality translation during training, we use an identity loss and we alter the traditional adversarial and cycle consistency losses to create an anatomically coherent adversarial loss and frequency cycle consistency losses.

Anatomically Coherent Adversarial Loss. The goal of the adversarial loss is to match the generated images to the corresponding real distribution. Inspired by the idea of conditional GANs [22] and pix2pix [14], we modify the adversarial loss to include as input the segmentation of the anatomical regions of interest. For the networks $\mathbb{G}_H$ and $\mathbb{D}_H$ our anatomically coherent adversarial loss is defined as:

$$\begin{aligned}\mathcal{L}_{\text{adv}}\left(\mathbb{G}_H, \mathbb{D}_H\right) = &\ \mathbb{E}_{h \sim p_{\text{data}}(h)}\left[\log \mathbb{D}_H(h, s_h)\right] \\ &+ \mathbb{E}_{l \sim p_{\text{data}}(l)}\left[\log\left(1 - \mathbb{D}_H(\mathbb{G}_H(l), s_l)\right)\right]\end{aligned} \tag{1}$$

By adding the segmentation as an input to the discriminator, we make sure that the networks learn the underlying relationship between the anatomical regions of interest and the structures in the generated image. Furthermore, the segmentation is not necessary at test time, since we only use the generator. In the final objective function, we also consider the adversarial loss for $\mathbb{G}_L$ and $\mathbb{D}_L$.

Frequency Cycle Consistency. The cycle consistency loss [30] ensures a direct one-to-one mapping from an image of one domain to another. However, the cycle consistency constraint is a pixel-wise L1 norm between the original image and the reconstruction, which enforces the output to have similar intensities. Yet, during the process of quality enhancement, it is more useful to think of the image in terms of frequency rather than intensity [10]. Low frequencies contain the structural information of an image, while high frequencies contain the fine details. With this concept in mind, we create two types of frequency consistency losses enforcing our training scheme to improve quality enhancement.

During quality translation, we aim to preserve the structural information present in the low frequencies of the original image. To extract low frequencies, we pass the images through a Gaussian pyramid [24] ϕ at K scales, then compute the L1 norm between the structural information of the original and the generated image (Eq. 2). We also want our generators to transfer image details of the corresponding quality in the form of high frequencies. Therefore, we obtain those frequencies through a Laplacian pyramid [24] γ at K scales and calculate the L1 norm between the high frequencies of the original image and the high frequencies of the reconstruction (Eq. 3). The loss concept is better illustrated in Fig. 1.

$$\mathcal{L}_{\text{lf}}\left(\mathbb{G}_H\right) = \sum_{k=1}^{K} \left\|\phi_k(l) - \phi_k\left(\mathbb{G}_H(l)\right)\right\|_1 \tag{2}$$

$$\mathcal{L}_{\text{hf}}\left(\mathbb{G}_H, \mathbb{G}_L\right) = \sum_{k=1}^{K} \|\gamma_k(l) - \gamma_k\left(\mathbb{G}_L(\mathbb{G}_H(l))\right)\|_1 \tag{3}$$

Identity Loss. The identity loss introduces a new constraint ensuring that the generator does not modify images from the same domain. This loss is particularly useful for quality enhancement in real clinical applications. In practice, we do not have a quality label but still we would like to transform all images to high quality without modifying the image if it already has a high quality.

$$\mathcal{L}_{\text{idt}} = \|h - \mathbb{G}_H(h)\|_1 \tag{4}$$

Overall Loss. For simplicity, we show only one of the pathways in the loss formulations, but our overall loss is defined as the weighted sum of the losses in both pathways $H \rightarrow L$ and $L \rightarrow H$, where each λ represents the relative importance of each loss function in the system: $\mathcal{L}_{\text{UltraGAN}} = \lambda_{\text{adv}}\mathcal{L}_{\text{adv}} + \lambda_{\text{lf}}\mathcal{L}_{\text{lf}} + \lambda_{\text{hf}}\mathcal{L}_{\text{hf}} + \lambda_{\text{idt}}\mathcal{L}_{\text{idt}}$.

3 Experiments

3.1 Dataset

To validate our method, we use the publicly available "Cardiac Acquisitions for Multi-structure Ultrasound Segmentation" (CAMUS) dataset [19]. The CAMUS dataset contains 2D Ultrasound images and multi-structure segmentations of 500 patients. This dataset is particularly relevant because it includes ultrasound images from three different qualities labeled by expert physicians. Besides, CAMUS includes patients with different left ventricle ejection fraction, making it a realistic problem with healthy and pathological subjects. The task in the CAMUS dataset is to segment the left ventricular endocardium (LV_{Endo}), left ventricular epicardium (LV_{Epi}) and left atrium (LA) in two chamber (2CH) and four chamber (4CH) views for End of Diastole (ED) and End of Systole (ES).

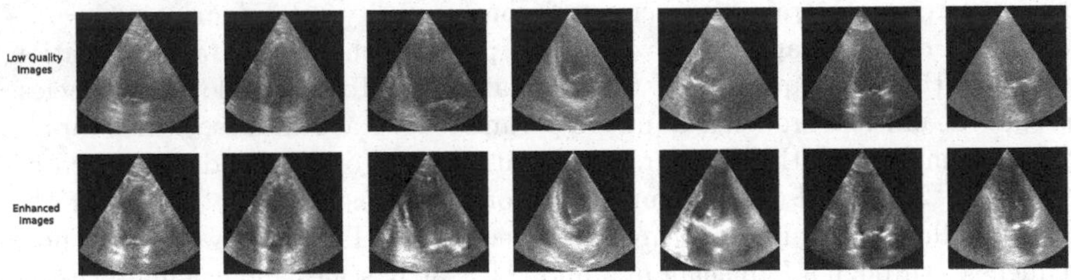

Fig. 2. Qualitative comparison of the low-quality and enhanced images using UltraGAN. Our method is able to enhance ultrasound images, improving the interpretability of the heart structures regardless of the view.

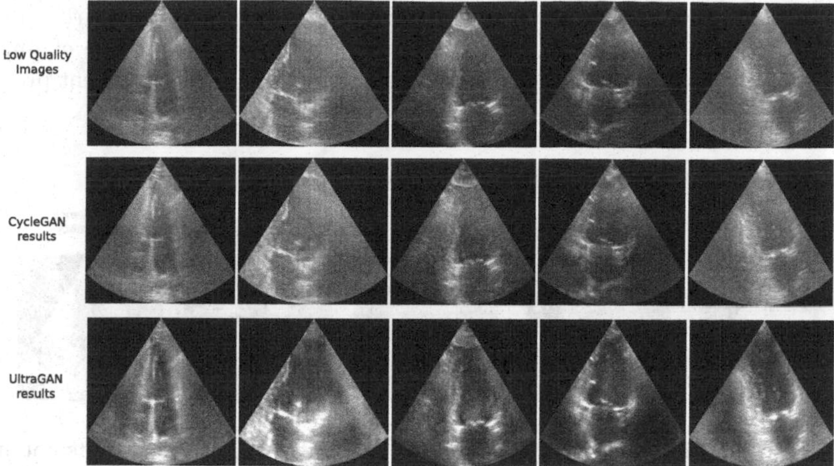

Fig. 3. Qualitative comparison between CycleGAN result and UltraGAN. The images generated by CycleGAN are perceptually similar to the original Low quality images. In contrast, images enhanced by UltraGAN show a clear difference between anatomical structures.

3.2 Experimental Setup

We first assess the importance of each of the components of our method in the task of quality enhancement. For simplicity, we divide the ultrasound images into high and low quality, classifying the medium quality images as low quality during our enhancement process. We train our method with 80% of the images and evaluate on the remaining images. We enhance the training data with three variants of our system: without the anatomically coherent adversarial loss, without the frequency cycle consistency losses, and with the original CycleGAN losses. Nevertheless, the evaluation of image quality in an unpaired setup is a subjective process and performing perceptual studies would require an extensive effort by expert physicians. However, we take advantage of the publicly available segmentation masks to provide multi-structure segmentation as a down-stream quantitative metric, in which the right global anatomical structure is required for a good performance.

As a baseline, we first train a simple U-Net model [27] using the standard 10 fold cross-validation split of the CAMUS dataset. Then, we use UltraGAN to enhance the quality of all the training images and train the same U-Net with the original images as well as the enhanced augmentation.

4 Results

4.1 Image Enhancement

The enhancement in image quality provided by UltraGAN is noticeable even for untrained eyes. Figure 2 shows the comparison between the low-quality images and the enhanced images we produce. In the enhanced images, the heart's chambers are recognizable and their boundaries are easy to identify. The examples illustrate the preservation of anatomical consistency in the enhancement process for both 2CH and 4CH views.

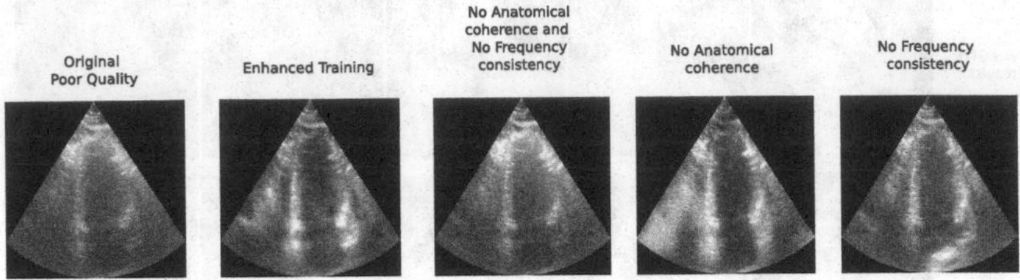

Fig. 4. Ablation examples of our enhancement method. We show the results obtained for every stage of the generation.

Furthermore, in Figs. 4 and 3 we demonstrate that the training choices for our method improve over the baseline, in which we just consider the standard loss function. We compare the original image against our enhancement showing that there is a better definition of the structure of the heart with more defined walls. We also compare the difference between having just the anatomical coherence or the frequency consistency. The images enhanced using merely frequency consistency maintain finer details, yet the system tends to hallucinate high frequencies in the left part of the image. Conversely, considering just the anatomical coherence, the structure is preserved but there is not a well definition of heart regions. Overall, with UltraGAN we are able to create an image quality enhancement that takes into account frequency and structural information.

4.2 Multi-structure Segmentation

In Fig. 5 we show that the segmentations obtained by using the standard data have artifacts, while training with UltraGAN-enhanced images improves the resulting segmentation. Also, for quantitative results, Table 1 shows the Dice

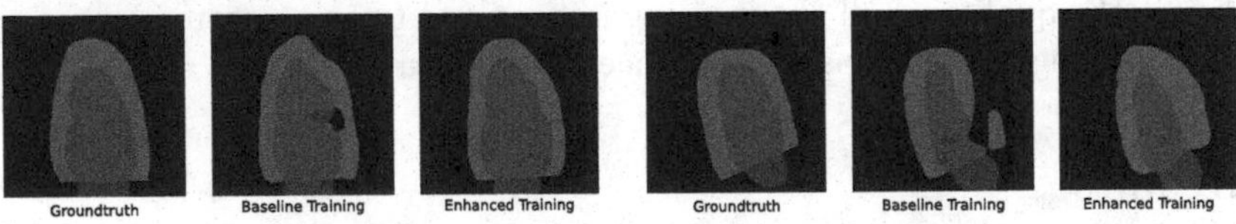

Fig. 5. Qualitative results for heart segmentation in the CAMUS dataset by using our enhanced images as data augmentation in the training stage. We present two different test examples showing the groundtruth (columns 1 and 4), the baseline results (columns 2 and 5) and the improved segmentation (columns 3 and 6).

Table 1. Segmentation results for 10-fold cross-validation set comparing standard training vs training with quality enhancement.

Method	High (%)			Medium (%)			Low (%)		
	LV_{Endo}	LV_{Epi}	LA	LV_{Endo}	LV_{Epi}	LA	LV_{Endo}	LV_{Epi}	LA
Baseline	93.07	86.61	88.99	92.02	85.32	88.13	90.76	83.10	87.52
Our method	**93.78**	**87.38**	**89.48**	**92.66**	**86.20**	**88.38**	**91.55**	**83.75**	**87.84**

Table 2. Segmentation results for 10-fold cross-validation comparing the state-of-the-art vs our quality enhanced training.

Image quality	Method	ED (%)		ES (%)	
		LV_{Endo}	LV_{Epi}	LV_{Endo}	LV_{Epi}
High	Ours	$\mathbf{94.40 \pm 0.7}$	86.54 ± 1.2	$\mathbf{92.04 \pm 1.1}$	87.05 ± 1.4
+ Medium	Leclerc *et al.*	93.90 ± 4.3	$\mathbf{95.40 \pm 2.3}$	91.60 ± 6.1	$\mathbf{94.50 \pm 3.9}$
Low	Ours	$\mathbf{93.00 \pm 1.1}$	83.57 ± 1.9	$\mathbf{90.10 \pm 1.3}$	83.93 ± 2.7
	Leclerc *et al.*	92.10 ± 3.7	$\mathbf{94.70 \pm 2.3}$	89.80 ± 5.7	$\mathbf{93.67 \pm 3.2}$

Scores for this experiment. Here we find evidence of the previous results, showing that for each of the structures present in the ultrasound image, augmenting the training data with UltraGAN improves the segmentation results. This improvement is also consistent across all of the image qualities, suggesting that the baseline with enhanced training data preserves correctly the anatomical structures present in the ultrasound images. We also evaluate separately the segmentation of our enhanced images in a subset of the CAMUS dataset consisting of patients at pathological risk with a left ventricle ejection fraction lower than 45%. We find that for pathological cases, the average Dice score (89.5) is as good as for healthy patients (89.7). Thanks to the global consistency enforced by the other heart structures, UltraGAN is able to extract accurately atypical left ventricles.

Table 2 shows the comparison between the state-of-the-art method in the CAMUS dataset and our quality enhanced method for the High+Medium and Low qualities in the 10 fold cross-validation sets. We do not include the comparison for Left atrium segmentation since the authors do not report their performance on that class. [19] uses a modified U-Net network to achieve the results. Here we demonstrate that, even with a simpler network with less amount of parameters, by enhancing the quality of the training images we are able to outperform state-of-the-art approaches in left ventricular endocardium segmentation, and obtain competitive results in left ventricular epicardium segmentation. Thus, demonstrating that the inclusion of quality enhanced images during training can benefit a model's generalization.

5 Conclusions

We present UltraGAN, a Generative Adversarial Network designed for quality enhancement of ultrasound images. We achieve image enhancement of 2D echocardiography images without compromising the anatomical structures. By using multi-structure segmentation as a downstream task we demonstrate that augmenting the training data with enhanced images improves the segmentation results. We expect UltraGAN to be useful in other ultrasound problems to push forward automated ultrasound analysis.

Acknowledgements. The present study is funded by MinCiencias, contract number 853-2019 project ID# 120484267276.

References

1. Abdi, A.H., Jafari, M.H., Fels, S., Tsang, T., Abolmaesumi, P.: A study into echocardiography view conversion. In: Workshop of Medical Imaging Meets NeurIPS (2019)
2. Abdi, A.H., Tsang, T., Abolmaesumi, P.: GAN-enhanced conditional echocardiogram generation. In: Workshop of Medical Imaging Meets NeurIPS (2019)
3. Abhishek, K., Hamarneh, G.: Mask2Lesion: mask-constrained adversarial skin lesion image synthesis. In: Burgos, N., Gooya, A., Svoboda, D. (eds.) Simulation and Synthesis in Medical Imaging, pp. 71–80. Springer International Publishing, Cham (2019). https://doi.org/10.1007/978-3-030-32778-1_8
4. Armstrong, A.C., et al.: Quality control and reproducibility in M-mode, two-dimensional, and speckle tracking echocardiography acquisition and analysis: the cardia study, year 25 examination experience. Echocardiography **32**(8), 1233–1240 (2015)
5. Cahalan, M.K., et al.: American society of echocardiography and society of cardiovascular anesthesiologists task force guidelines for training in perioperative echocardiography. J. Am. Soc. Echocardiogr. **15**(6), 647–652 (2002)
6. Chen, L., Bentley, P., Mori, K., Misawa, K., Fujiwara, M., Rueckert, D.: Intelligent image synthesis to attack a segmentation CNN using adversarial learning. In: Burgos, N., Gooya, A., Svoboda, D. (eds.) Simulation and Synthesis in Medical Imaging, pp. 90–99. Springer International Publishing, Cham (2019). https://doi.org/10.1007/978-3-030-32778-1_10
7. Chen, Y.S., Wang, Y.C., Kao, M.H., Chuang, Y.Y.: Deep photo enhancer: unpaired learning for image enhancement from photographs with GANs. In: Proceedings of the IEEE Conference on Computer Vision and Pattern Recognition, pp. 6306–6314 (2018)
8. Deshpande, A., Lu, J., Yeh, M.C., Jin Chong, M., Forsyth, D.: Learning diverse image colorization. In: Proceedings of the IEEE Conference on Computer Vision and Pattern Recognition, pp. 6837–6845 (2017)
9. Duarte-Salazar, C.A., Castro-Ospina, A.E., Becerra, M.A., Delgado-Trejos, E.: Speckle noise reduction in ultrasound images for improving the metrological evaluation of biomedical applications: an overview. IEEE Access **8**, 15983–15999 (2020)
10. Fritsche, M., Gu, S., Timofte, R.: Frequency separation for real-world super-resolution. In: ICCV Workshop (2019)

11. Gardin, J.M., et al.: Recommendations for a standardized report for adult transthoracic echocardiography: a report from the American society of echocardiography's nomenclature and standards committee and task force for a standardized echocardiography report. J. Am. Soc. Echocardiogr. **15**(3), 275–290 (2002)
12. Goodfellow, I., et al.: Generative adversarial nets. In: Advances in Neural Information Processing Systems, pp. 2672–2680 (2014)
13. Gottdiener, J.S., et al.: American society of echocardiography recommendations for use of echocardiography in clinical trials: a report from the American society of echocardiography's guidelines and standards committee and the task force on echocardiography in clinical trials. J. Am. Soc. Echocardiogr. **17**(10), 1086–1119 (2004)
14. Isola, P., Zhu, J.Y., Zhou, T., Efros, A.A.: Image-to-image translation with conditional adversarial networks. In: 2017 IEEE Conference on Computer Vision and Pattern Recognition (CVPR) (2017)
15. Jafari, M.H., et al.: Semi-supervised learning for cardiac left ventricle segmentation using conditional deep generative models as prior. In: 2019 IEEE 16th International Symposium on Biomedical Imaging (ISBI 2019), pp. 649–652. IEEE (2019)
16. Jafari, M.H., et al.: Cardiac point-of-care to cart-based ultrasound translation using constrained cyclegan. Int, J. Comput. Assist. Radiol. Surg. **15**(5), 877–886 (2020). https://doi.org/10.1007/s11548-020-02141-y
17. Jafari, M.H., et al.: Echocardiography segmentation by quality translation using anatomically constrained CycleGAN. In: Shen, D., et al. (eds.) MICCAI 2019. LNCS, vol. 11768, pp. 655–663. Springer, Cham (2019). https://doi.org/10.1007/978-3-030-32254-0_73
18. Lartaud, P.-J., Rouchaud, A., Rouet, J.-M., Nempont, O., Boussel, L.: Spectral CT based training dataset generation and augmentation for conventional CT vascular segmentation. In: Shen, D., et al. (eds.) MICCAI 2019. LNCS, vol. 11765, pp. 768–775. Springer, Cham (2019). https://doi.org/10.1007/978-3-030-32245-8_85
19. Leclerc, S., et al.: Deep learning for segmentation using an open large-scale dataset in 2D echocardiography. IEEE Trans. Med. Imaging **38**(9), 2198–2210 (2019)
20. Liao, Z., et al.: Echocardiography view classification using quality transfer star generative adversarial networks. In: Shen, D., et al. (eds.) MICCAI 2019. LNCS, vol. 11765, pp. 687–695. Springer, Cham (2019). https://doi.org/10.1007/978-3-030-32245-8_76
21. Mazaheri, S., et al.: Echocardiography image segmentation: a survey. In: 2013 International Conference on Advanced Computer Science Applications and Technologies, pp. 327–332. IEEE (2013)
22. Mirza, M., Osindero, S.: Conditional generative adversarial nets. arXiv preprint arXiv:1411.1784 (2014)
23. Ng, A., Swanevelder, J.: Resolution in ultrasound imaging. Continuing Educ. Anaesth. Crit. Care Pain **11**(5), 186–192 (2011)
24. Oliva, A., Torralba, A., Schyns, P.G.: Hybrid images. ACM Trans. Graph. (TOG) **25**(3), 527–532 (2006)
25. Ortiz, S.H.C., Chiu, T., Fox, M.D.: Ultrasound image enhancement: a review. Biomed. Signal Process. Control **7**(5), 419–428 (2012)
26. Romero, A., Arbeláez, P., Van Gool, L., Timofte, R.: SMIT: stochastic multi-label image-to-image translation. In: Proceedings of the IEEE International Conference on Computer Vision Workshops (2019)
27. Ronneberger, O., Fischer, P., Brox, T.: U-Net: convolutional networks for biomedical image segmentation. In: Navab, N., Hornegger, J., Wells, W.M., Frangi, A.F.

(eds.) MICCAI 2015. LNCS, vol. 9351, pp. 234–241. Springer, Cham (2015). https://doi.org/10.1007/978-3-319-24574-4_28
28. Wang, X., Chan, K.C., Yu, K., Dong, C., Loy, C.C.: EDVR: video restoration with enhanced deformable convolutional networks. In: The IEEE Conference on Computer Vision and Pattern Recognition (CVPR) Workshops, June 2019
29. Yang, H., et al.: Unpaired brain MR-to-CT synthesis using a structure-constrained CycleGAN. In: Stoyanov, D., et al. (eds.) DLMIA/ML-CDS -2018. LNCS, vol. 11045, pp. 174–182. Springer, Cham (2018). https://doi.org/10.1007/978-3-030-00889-5_20
30. Zhu, J.Y., Park, T., Isola, P., Efros, A.A.: Unpaired image-to-image translation using cycle-consistent adversarial networks. In: 2017 IEEE International Conference on Computer Vision (ICCV) (2017)

Improving Endoscopic Decision Support Systems by Translating Between Imaging Modalities

Georg Wimmer[1], Michael Gadermayr[2(✉)], Andreas Vécsei[3], and Andreas Uhl[1]

[1] Department of Computer Science, University of Salzburg, Salzburg, Austria
[2] Salzburg University of Applied Sciences, Salzburg, Austria
michael.gadermayr@fh-salzburg.ac.at
[3] St. Anna Children's Hospital, Vienna, Austria

Abstract. Novel imaging technologies raise many questions concerning the adaptation of computer-aided decision support systems. Classification models either need to be adapted or even newly trained from scratch to exploit the full potential of enhanced techniques. Both options typically require the acquisition of new labeled training data. In this work we investigate the applicability of image-to-image translation to endoscopic images captured with different imaging modalities, namely conventional white-light and narrow-band imaging. In a study on computer-aided celiac disease diagnosis, we explore whether image-to-image translation is capable of effectively performing the translation between the domains. We investigate if models can be trained on virtual (or a mixture of virtual and real) samples to improve overall accuracy in a setting with limited labeled training data. Finally, we also ask whether a translation of the images to be classified is capable of improving accuracy by exploiting imaging characteristics of the new domain.

Keywords: Image-to-image translation · Generative adversarial networks · Cycle-GAN · Endoscopy · Narrow-band imaging · Data augmentation

1 Motivation

Optimum visibility of the crucial visual features for a reliable diagnosis is an important criterion for imaging devices in endoscopy. Narrow-band imaging (NBI) [4,13] is an imaging technique enhancing the details of the surface of the mucosa. A filter is electronically activated letting pass ambient light of wavelengths of 415 (blue) and 540 nm (green) only. Conventional endoscopes illuminate the mucosa with a broad visible light-spectrum. Strong indication is provided that also computer-aided decision support systems show improved accuracy if being applied to NBI data instead of conventional white-light imaging (WLI) data [14]. Due to the clearly different image characteristics of the two

G. Wimmer and M. Gadermayr—Equal Contributions.

N. Burgos et al. (Eds.): SASHIMI 2020, LNCS 12417, pp. 131–141, 2020.
https://doi.org/10.1007/978-3-030-59520-3_14

imaging modalities NBI and WLI, classification models trained for one of the modalities cannot be directly applied to the other modality without clearly loosing accuracy [14]. The straight-forward approach is to train an individual model for NBI and one for WLI. However, convolutional neural networks (CNNs), which are the state-of-the-art in endoscopic image classification, require large amounts of labeled data showing high visual quality. As there is typically a lack of such data, a major question is, if data from both domains can be used to train either individual models or one generic model for both domains. For that purpose, typically domain adaptation methods are applied. Conventional domain adaptation integrates the adaptation of the data in the image analysis pipeline. For that reason, domain adaptation is typically linked to the final image analysis approach. Image-to-image translation can be interpreted as domain adaptation on image level [7] and is therefore completely independent of the final task as the image itself is converted. Image-to-image translation gained popularity during the last years generating highly attractive and realistic output [6,16]. While many approaches require image pairs for training the image transformation models, generative adversarial networks (GANs) [6], making use of the so-called cycle-consistency loss (cycle-GANs), are free from this restriction. They do not require any image pairs for domain adaptation and are completely independent of the underlying image analysis (e.g. classification) task. Finally, they also improve interpretability of the adaptation process by allowing visual investigation of the generated virtual (or "fake") images. For the purpose of domain adapatation, cycle-GANs were effectively used in the field of digital pathology [5] and radiology for translating between the modalities, such as MRI and CT [15].

In the experimental evaluation, we consider the applicability to computer aided celiac disease (CD) diagnosis. CD is a multisystemic immune-mediated disease, which is associated with considerable morbidity and mortality [1]. Endoscopy in combination with biopsy is currently considered the gold standard for the diagnosis of CD. Computer-aided decision support systems have potential to improve the whole diagnostic work-up, by saving costs, time and manpower and at the same time increase the safety of the procedure [14]. A motivation for such a system is furthermore given as the inter-observer variability is reported to be high [1,2].

Contribution

In this work, we investigate whether decision support systems for celiac disease diagnosis can be improved by making use of image-to-image translation. Specifically, we perform translations between images captured using the NBI technology and conventional WLI. Several experiments are performed in order to answer the following questions:

- Q0: can image-to-image translation be effectively applied to endoscopic images?
- Q1: can virtual images be used to increase the training data set and finally improve classification accuracy?

- Q2: how do virtual images perform when being used during testing? Do virtual images in the more discriminative NBI domain maybe even show improved performance compared to real WLI samples?

2 Methods

For answering these questions, several pipelines (Fig. 1) combining an image translation model (Sect. 2.1) with classification models (Sect. 2.2) are developed and evaluated. Particularly, different combinations of imaging modalities are employed for training and testing.

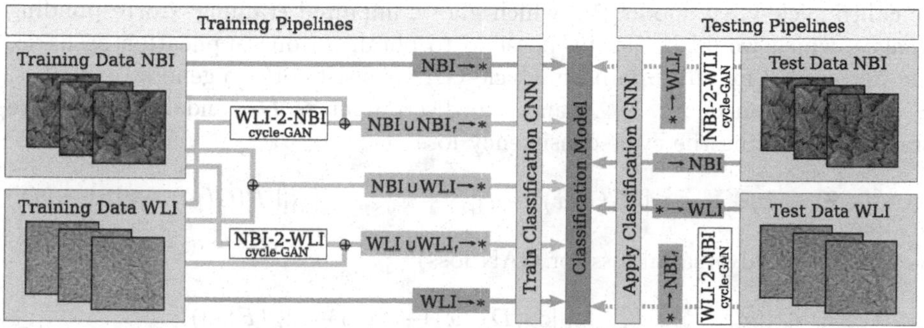

Fig. 1. Overview of the performed experiments: training is conducted with data of a single (e.g. NBI→ $*$) and both domains without (NBI ∪ WLI→ $*$) and with adaptation (e.g. NBI ∪ NBI_f→ $*$). For testing either real samples (e.g. $*$ → NBI) or translated images are applied (e.g. $* \rightarrow NBI_f$).

As a baseline, experiments using training and testing data from the same original imaging modality are performed (e.g. Fig. 1, NBI → NBI). To answer Q1 ("Can virtual images be used to increase the training data set and finally improve classification accuracy?"), we conduct the following experiments which are compared with the baseline: NBI and WLI images are combined for training the classification model (NBI ∪ WLI → $*$) which is applied to both data sets separately for testing. Additionally, NBI and virtual NBI (NBI ∪ NBI_f) samples (i.e. WLI samples which were translated to NBI) are used to train a model for NBI. The same is applied to WLI samples using real WLI and virtual WLI (WLI ∪ WLI_f) samples to train a model to test WLI data.

To answer Q2 ("How do virtual images perform when being used during testing?"), the following experiments are performed and compared with the baseline. First, we assess the effect if virtual samples are used for training only (NBI_f → $*$). In another experiment, real and virtual (e.g. NBI ∪ NBI_f) samples are merged for training to enlarge the training corpus. Testing is performed either on original ($*$→ NBI) or virtual image samples showing the same domain ($*$→ NBI_f). In that way, we want to find out if the virtual data is similarly suited for testing as compared to the original endoscopic image data. This is especially promising in the case of translating WLI to NBI, since NBI is supposed to improve the results of automated diagnosis systems compared to WLI [14].

The question Q0 ("Does image-to-image translation work for endoscopic images?") is finally answered by means of visual inspection of the generated translated images and by considering the quantitative experiments mentioned above.

2.1 Image Translation Network

For performing translation between the NBI and the WLI data (and vice versa), we employ a generative adversarial network (GAN). Specifically we utilize the so-called cycle-GAN model [16] which allows unpaired training. Corresponding images, which would be hardly possible to obtain in our application scenario, are not needed for this approach. Cycle-GAN consists of two generator models, $F : X \rightarrow Y$ and $G : Y \rightarrow X$ and two discriminators D_X and D_Y which are trained optimizing the cycle-consistency loss

$$\mathscr{L}_c = \mathbb{E}_{x \sim p_{data}(x)}[||G(F(x)) - x||_1] + \mathbb{E}_{y \sim p_{data}(y)}[||F(G(y)) - y||_1] \tag{1}$$

as well as the adversarial loss (or GAN loss)

$$\begin{aligned} \mathscr{L}_d = & \ \mathbb{E}_{x \sim p_{data}(x)}[\log(D_X(x)) + \log(1 - D_Y(F(x)))] \\ & + \mathbb{E}_{y \sim p_{data}(y)}[\log(1 - D_X(G(y))) + \log(D_Y(y))] \,. \end{aligned} \tag{2}$$

F and G are trained to fool the discriminators and thereby generate fake images that look similar to real images, while D_X and D_Y are optimized to distinguish between translated and real samples for both domains individually. While the generators aim to minimize this adversarial loss, the discriminators aim to maximize it.

To train the image-translation model, for both data sets, all original endoscopic images are padded to fit a common size (768×768). With these patches, a cycle-GAN is trained [16] (with equal weights $w_{cyc} = 1, w_{GAN} = 1$). We do not use the optional identity loss $\mathscr{L}_{id}$. As generator network, we employ the U-Net [11]. Apart from that, the standard configuration based on the patch-wise CNN discriminator is utilized [16])[1]. Training is performed for 1000 epochs. The initial learning rate is set to 10^{-5}. Random flipping and rotations (0, 90, 180, 270°) are applied for data augmentation.

[1] We use the provided PyTorch reference implementation [16].

2.2 Classification Models

We investigate three CNNs for performing classification of celiac disease based on endoscopic image patches. The networks consist of AlexNet [8], VGG-f [3] and VGG-16 [12]. These models are chosen as they exhibit high potential for endoscopic image analysis [10,14]. The networks are trained on patches of a size of 256 $\times$ 256 pixels (the original image patches of the employed data sets are upscaled from 128×128 to 256×256 pixels using bicubic interpolation [10]). To prevent any bias, all networks are randomly initialized and trained from scratch. The size of the last fully-connected layer is adapted to the considered two-classes classification scheme. The last fully connected layer is acting as soft-max classifier and computes the training loss (log-loss). Stochastic gradient descent with weight decay $\lambda = 0.0005$ and momentum $\mu = 0.9$ is employed for the optimization. Training is performed on batches of 128 images each, which are randomly selected (independently for each iteration) from the training data and subsequently augmented. For augmentation, random cropping (224×224 pixel patches for VGG-f and VGG-16 and 227 $\times$ 227 pixel patches for Alex-net), horizontal flipping and random rotations with multiples of 90° are performed.

2.3 Experimental Details and Data Sets

In this work we consider a two-class classification and differentiate between healthy mucosa and mucosa affected by CD using images gathered by NBI as well as WLI (both in combination with the modified immersion technique [14]). During endoscopy, images were taken either from both domains or only from WLI or NBI. The endoscopic images were acquired from 353 patients overall. Classification of CD is aggravated by the patchy-distribution of CD and by the fact that the abstinence of visible features can easily cause a wrong categorization. For that purpose, we utilize manually extracted image patches showing distinctive regions-of-interest for training the classification models. For training the image translation model, the complete full-resolution original images are applied. The WLI image data set consists of 1045 image patches (587 healthy images and 458 affected by CD) from 313 patients and the NBI database consists of 616 patches (399 healthy images and 217 affected by celiac disease) from 123 patients. Figure 2 shows example patches showing healthy mucosa and mucosa affected by CD, for both considered imaging modalities individually.

In all experiments, five-fold cross-validation is performed to achieve a stable estimation of the generalization error, where each of the five folds consists of the images of approx. 20% of all patients. To avoid bias, we ensure that all images of one patient are in the same fold. By means of the McNemar test [9], we assess the statistical significance of the obtained improvements.

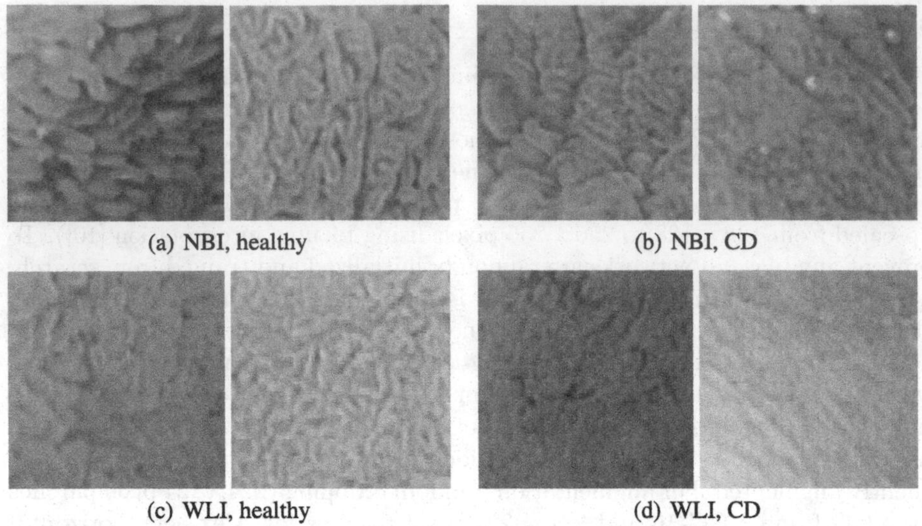

Fig. 2. Example images for the two classes healthy and celiac disease (CD) using NBI as well as WLI endoscopy.

3 Results

In Table 1, the overall classification rates for the individual experiments are shown. For each CNN architecture and each sub-experiment (sub-experiments are separated by vertical lines in Table 1), the accuracy of the best performing combination of training and test data set is given in boldface numbers. Apart from the first sub-experiment, the scores within each sub-experiment can be directly compared as the underlying image content is the same for the considered testing data. A direct unbiased numeric comparison between the scores for the NBI and WLI data set is not possible since the data sets show different mucosal areas. Consequently, the baseline comparison between NBI and WLI does not directly answer the question for the more suitable imaging modality for computer aided diagnosis.

When comparing the outcomes of the standard setting with the same domain for training and testing (baseline results), we notice that the accuracies are slightly higher (+0.6% on average) in case of WLI image data. However, here we need to remark that the overall data set (and hence also the data available for training) is larger (WLI: 1045 patches; NBI: 616 patches). Additionally, the testing data does not capture the same mucosal regions.

In the two sub-experiments to answer Q1, the effect of virtual training data is assessed. Using only NBI images as training data for testing NBI samples showed higher scores (+5.4% on average) than mixing NBI and WLI data (without any domain adaptation) for training. Combining NBI data with virtual NBI_f samples for training on slightly improved the accuracies (+0.3% on average) compared to using NBI data only for training. Using only WLI data for training when

Table 1. Mean classification accuracies for the CNNs using different combinations of training and test data sets. WLI_f refers to samples originally captured under NBI which were translated to WLI. NBI_f refers to WLI samples translated to NBI. The asterisk (*) indicates statistical significant improvements ($p < 0.05$).

Training		Test	VGG-f	Alex-net	VGG-16	∅
Baseline for WLI and NBI data						
WLI	→	WLI	87.4(2.4)	87.6(1.5)	85.9(2.5)	87.0
NBI	→	NBI	87.6(3.8)	88.1(4.1)	83.5(3.4)	86.4
Q1: Testing NBI samples with varying training data						
NBI	→	NBI	87.6(3.8)	**88.1**(4.1)	83.5(3.4)	86.4
NBI ∪ WLI	→	NBI	81.3(3.9)	81.5(6.0)	80.3(5.3)	81.0
NBI ∪ NBI_f	→	NBI	**88.3**(2.6)	87.0(2.4)	**84.9**(5.3)	**86.7**
Q1: Testing WLI samples with varying training data						
WLI	→	WLI	87.4(2.4) ⌉	87.6(1.5) ⌉	85.9(2.5)	87.0
NBI ∪ WLI	→	WLI	87.3(2.4) | *	88.2(1.3) | *	86.3(2.2)	87.3
WLI_f ∪ WLI	→	WLI	**89.8**(1.6) ⌋	**89.6**(1.7) ⌋	**87.5**(3.3)	**89.0**
Q2: Real WLI vs. virtual NBI data for testing						
WLI	→	WLI	87.4(2.4) ⌉ *	87.6(1.5) ⌉ *	85.9(2.5)	87.0
WLI_f ∪ WLI	→	WLI	**89.8**(1.6) ⌋	**89.6**(1.7) ⌋	87.5(3.3)	**89.0**
NBI_f	→	NBI_f	88.9(2.2)	89.5(2.0)	**88.2**(3.1)	88.9
NBI ∪ NBI_f	→	NBI_f	88.8(2.5)	89.2(1.4)	86.6(2.4)	88.2
Q2: Real NBI vs. virtual WLI data for testing						
NBI	→	NBI	87.6(3.8)	**88.1**(4.1)	83.5(3.4)	86.4
NBI ∪ NBI_f	→	NBI	**88.3**(2.6)	87.0(2.4)	**84.9**(5.3)	**86.7**
WLI_f	→	WLI_f	82.9(2.9)	81.9(2.8)	78.5(2.7)	81.1
WLI ∪ WLI_f	→	WLI_f	85.6(5.6)	85.1(6.5)	83.5(6.7)	84.7

testing on WLI samples achieved the lowest scores. Additionally using NBI data for training, slightly improved the accuracies (+0.3% on average) but the clearly best results were achieved combining WLI images and virtual WLI_f images (+2.0% on average).

In the two sub-experiments to answer Q2, the effect of classifying translated samples instead of the original ones is assessed. Using samples of the WLI data set for classification (either the original samples or the samples translated to NBI_f), the best scores are achieved when classifying the original WLI data using WLI and virtual WLI_f data for training. Training and testing on WLI data

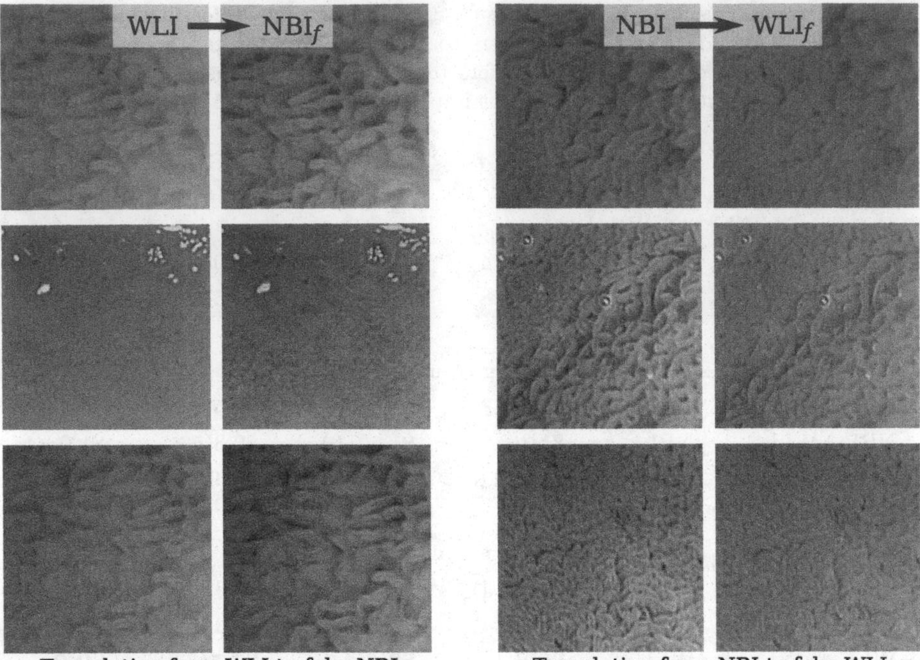

Fig. 3. Example images before and after the image-to-image translation process. The figures on the left show original WLI samples and the original WLI samples translated to NBI (NBI$_f$) and the figures to the right show original NBI samples and the original NBI samples translated to WLI (WLI$_f$).

translated to NBI (NBI$_f$) shows similar performance (−0.1% on average). For the classification of samples from the NBI data set (either the original samples or the samples translated to WLI$_f$), the best accuracies are achieved when classifying the original NBI data (83.5–88.3% on average). When classifying NBI data translated to WLI, accuracies drop to between 78.5 and 85.6%. Example translated "fake" images and the corresponding original "real" samples are provided in Fig. 3 for both translation directions.

4 Discussion

In this paper, we investigated whether decision support systems for celiac disease diagnosis can be improved by making use of image translation. Experiments were performed to answer three questions posed in Sect. 1.

We first discuss whether image-to-image translation is in principle applicable for the considered endoscopic image data (Q0). Regarding the example fake images (see Fig. 3) and the quantitative final classification scores, we can conclude that image translation is able to generate useful virtual images in both directions. In general, the images look realistic. In some cases, we notice that the translation process generates images with a high similarity to the original domain. This is supposed to be due to the fact that the data sets also contain images which look like hybrids between NBI and WLI depending on the illumination quality and the perspective of the camera towards the mucosal wall. There are huge differences with respect to illumination and perspective in the endoscopic image data. Regarding the overall classification rates, however, we notice that translated images, either used alone or in combination with original images for training, generally show high accuracy (especially WLI data converted to NBI).

Next, we asked whether virtual images should be used to enlarge the training data set to finally improve classification accuracy in settings with limited annotated training data (Q1). We can summarize that this is definitely the case. We notice increased accuracies in five out of six settings when combining fake and original samples for classification. This effect is observed independent of the direction of translation, i.e. improvements are obtained for NBI-to-WLI translation as well as for WLI-to-NBI translation. The accuracies also clearly increase when combining e.g. NBI and NBI_f for training instead of simply combining the NBI and the WLI data set when testing NBI samples. For that reason, we can conclude that the effect of the domain shift between NBI and WLI data is clearly reduced when performing image translation.

Finally, we asked how domain transferred images perform when being used for classification instead of the original images (Q2)? To answer this question, a differentiation for the two data sets needs to be applied. A translation of test data to the WLI domain (WL_f) generally shows decreased rates. As already indicated in [14], this is probably because the WLI modality is generally less suited for a computer-based classification than NBI. Even a perfect translation to a less suited modality obviously cannot exhibit better scores. Considering a translation to the NBI domain, we notice improvement if virtual samples are used for training and testing only compared to using original WLI data for training and testing. This also corroborates the assumption that NBI can be more effectively classified and that the translation process works really well. Interestingly, using original samples additionally to the translated ones does not show further improvements. Even though the translation in principle works well, it can be assumed that there is still some domain gap left between real and fake samples. Finally, it could be the case that the virtual NBI_f domain requires fewer samples for effectively training a classification network. This could be due to clearer highlighted features and/or decreased variability in the data.

5 Conclusion

We showed that image-to-image translation can be effectively applied to endoscopic images in order to change the imaging modality from WLI to NBI and vice versa. Evaluation exhibited that additionally employing virtual training data is capable of increasing the classification rates of CNNs. Converted images applied for testing also showed slightly increased accuracies if conversion is performed from WLI to the obviously more appropriate NBI domain. This study therefore also provides (further) indication that NBI is more suited than WLI for computer aided decision-support systems since the translation to the NBI domain clearly achieves better results than the translation to WLI. Finally, the obtained insights provide incentive to further increase distinctiveness by translating to special subdomains showing particularly high image quality in future work. A distinctiveness-contraint could also be incorporated into the GAN architecture.

Acknowledgement. This work was partially funded by the County of Salzburg under grant number FHS-2019-10-KIAMed.

References

1. Biagi, F., Corazza, G.: Mortality in celiac disease. Nat. Rev. Gastroenterol. Hepatol. **7**, 158–162 (2010)
2. Biagi, F., et al.: Video capsule endoscopy and histology for small-bowel mucosa evaluation: a comparison performed by blinded observers. Clin. Gastroenterol. Hepatol. **4**(8), 998–1003 (2006)
3. Chatfield, K., Simonyan, K., Vedaldi, A., Zisserman, A.: Return of the devil in the details: delving deep into convolutional nets. In: British Machine Vision Conference, BMVC 2014, Nottingham, UK, 1–5 September 2014
4. Emura, F., Saito, Y., Ikematsu, H.: Narrow-band imaging optical chromocolonoscopy: advantages and limitations. World J. Gastroenterol. **14**(31), 4867–4872 (2008)
5. Gadermayr, M., Gupta, L., Appel, V., Boor, P., Klinkhammer, B.M., Merhof, D.: Generative adversarial networks for facilitating stain-independent supervised and unsupervised segmentation: a study on kidney histology. IEEE Trans. Med. Imaging **38**(10), 2293–2302 (2019)
6. Isola, P., Zhu, J.Y., Zhou, T., Efros, A.A.: Image-to-image translation with conditional adversarial networks. In: Proceedings of the International Conference on Computer Vision and Pattern Recognition (CVPR 2017) (2017)
7. Kamnitsas, K., et al.: Unsupervised domain adaptation in brain lesion segmentation with adversarial networks. CoRR, http://arxiv.org/abs/1612.08894 (2016)
8. Krizhevsky, A., Sutskever, I., Hinton, G.E.: ImageNet classification with deep convolutional neural networks. In: Advances in Neural Information Processing Systems, vol. 25, pp. 1097–1105. Curran Associates Inc. (2012)
9. McNemar, Q.: Note on the sampling error of the difference between correlated proportions or percentages. Psychometrika **12**(2–3), 153–157 (1947)
10. Ribeiro, E., et al.: Exploring texture transfer learning for colonic polyp classification via convolutional neural networks. In: Proceedings of the 14th International IEEE Symposium on Biomedical Imaging (ISBI 2017), April 2017
11. Ronneberger, O., Fischer, P., Brox, T.: U-Net: convolutional networks for biomedical image segmentation. In: Navab, N., Hornegger, J., Wells, W.M., Frangi, A.F. (eds.) MICCAI 2015. LNCS, vol. 9351, pp. 234–241. Springer, Cham (2015). https://doi.org/10.1007/978-3-319-24574-4_28
12. Simonyan, K., Zisserman, A.: Very deep convolutional networks for large-scale image recognition. In: International Conference on Learning Representations (ICLR) (2015)

13. Valitutti, F., et al.: Narrow band imaging combined with water immersion technique in the diagnosis of celiac disease. Dig. Liver. Dis. **46**(12), 1099–1102 (2014)
14. Wimmer, G., et al.: Quest for the best endoscopic imaging modality for computer-assisted colonic polyp staging. World J. Gastroenterol. **25**(10), 1197–1209 (2019). https://doi.org/10.3748/wjg.v25.i10.1197
15. Wolterink, J.M., Dinkla, A.M., Savenije, M.H.F., Seevinck, P.R., van den Berg, C.A.T., Išgum, I.: Deep MR to CT synthesis using unpaired data. In: Tsaftaris, S.A., Gooya, A., Frangi, A.F., Prince, J.L. (eds.) SASHIMI 2017. LNCS, vol. 10557, pp. 14–23. Springer, Cham (2017). https://doi.org/10.1007/978-3-319-68127-6_2
16. Zhu, J.Y., Park, T., Isola, P., Efros, A.A.: Unpaired image-to-image translation using cycle-consistent adversarial networks. In: Proceedings of the International Conference on Computer Vision (ICCV 2017) (2017)

An Unsupervised Adversarial Learning Approach to Fundus Fluorescein Angiography Image Synthesis for Leakage Detection

Wanyue Li[1,2,3], Yi He[1,2], Jing Wang[1,2,3], Wen Kong[1,2,3], Yiwei Chen[1,2], and GuoHua Shi[1,2,3,4](✉)

[1] Suzhou Institute of Biomedical Engineering and Technology, Chinese Academy of Sciences, Suzhou, China
ghshi_lab@126.com

[2] Jiangsu Key Laboratory of Medical Optics, Suzhou, China

[3] University of Science and Technology of China, Hefei, China

[4] Center for Excellence in Brain Science and Intelligence Technology, Chinese Academy of Sciences, Shanghai, China

Abstract. Detecting the high-intensity retinal leakage in fundus fluorescein angiography (FA) image is a key step for retinal-related disease diagnosis and treatment. In this study, we proposed an unsupervised learning-based fluorescein leakage detecting method which can give the leakage detection results without the need for manual annotation. In this method, a model that can generate the normal-looking FA image from the input abnormal FA image is trained; and then the leakage can be detected by making the difference between the abnormal and generated normal image. The proposed method was validated on the publicly available datasets, and qualitatively and quantitatively compared with the state-of-the-art leakage detection methods. The comparison results indicate that the proposed method has higher accuracy in leakage detection, and can detect an image in a very short time (in 1 s), which has great potential significance for clinical diagnosis.

Keywords: Fundus fluorescein angiography · Unsupervised learning · Leakage detection

1 Introduction

Fundus fluorescein angiography (FA) is a commonly used retinal imaging modality that provides a map of retinal vascular structure and function by highlighting blockage of, and leakage from, retinal vessels [1, 2], which is regarded as the "gold standard" of retinal diseases diagnosis [3]. Detecting the high-intensity retinal leakage in angiography is a key step for clinicians to determine the activities and development of lesions in the retina, that enable decision-making for treatment and monitoring of disease activities. Current practical approaches for quantitative analysis of fluorescein leakage are usually delineated by trained graders [2]; this manually labeled method is time-consuming and labor-intensive, and usually introduces human errors. Therefore, an effective automated method for fluorescein leakage detection is warranted.

N. Burgos et al. (Eds.): SASHIMI 2020, LNCS 12417, pp. 142–152, 2020.
https://doi.org/10.1007/978-3-030-59520-3_15

Algorithms that tackle automated fluorescein leakage detection task can be mainly classified as either image intensity-based or supervised learning-based methods. The intensity-based method usually extracts and analyzes the intensity information to detect the leakage. In [4], the authors proposed a method to detect the leakage of macular edema in ocular vein occlusions. The authors suggested that any pixels with statistically high increments in gray level along the FA sequence close to the foveal center could be segmented as leakage; however, this method needs the manual detection of the foveal center, and cannot be regarded as a fully automated method. Rabbani et al. [5] use an active contour segmentation model to realize the fluorescein leakage detection for the subjects with diabetic macular edema. This method is designed to detect a local area (1500 μm radius circular region centered at the fovea) in the image and has a relatively low sensitivity (0.69) on 24 images. Zhao et al. [6] proposed a method to detect three types of leakage (large focal, punctate focal, and vessel segment leakage) on the images from malarial retinopathy eyes; this method extracts the intensity information to calculate the saliency map for the detection. However, this method may suffer when some non-leakage areas also have high intensities. To overcome this problem, they then proposed a new method [2] that uses intensity and compactness as two saliency cues to help more accurate detection of leakage. Although this method has relatively higher sensitivity and accuracy in fluorescein leakage detection, this kind of intensity-based method needs a relatively long time to detect an image (nearly 20 s). In terms of supervised learning-based method, Trucco et al. [7] and Tsai et al. [8] applied AdaBoost method to segment the leakage regions of FA images, however, these supervised learning methods need the training datasets derived from manual annotation, which makes the performance of the method is inherently dependent on the quality of the annotation. Moreover, the dataset is very important for supervised learning-based methods, and it is difficult to obtain large-scale data annotated by experienced graders.

Recently, with the development of generative adversarial networks (GANs) and unsupervised image translation techniques, unsupervised-learning-based lesion detection methods have also emerged. In [9], the authors proposed an abnormal-to-normal translation GAN to generate a normal-looking medical image based on its abnormal-looking counterpart, and the difference between the abnormal image and the generated normal image can guide the detection or segmentation of lesions. In [10], the authors proposed a model named fast AnoGAN (f-AnoGAN) applied to the anomaly detection of optical coherence tomography (OCT) retinal images, and this method can achieve nearly real-time anomaly detection. However, to the best of our knowledge, unsupervised learning-based methods have not been investigated for fluorescein leakage detection. Similarly, deep learning-based methods for fundus fluorescein leakage detection is still lacking in the current literature.

In this paper, inspired by the idea in [9, 11], we proposed an unsupervised model based on the cyclic generative adversarial network [12] (CycleGAN) to synthesize the normal-looking FA images based on the abnormal FA images, and then the difference between normal and abnormal FA images are used to help us detect leakages. This unsupervised method mainly has two advantages: one is that this method can give the leakage detection results without any either manual annotation or paired normal and abnormal images, this method only needs the normal and abnormal FA images and uses these data to learn how to generate normal-looking FA images. The other one is that the deep-learning-based method takes up a long time in the training stage, but it performs quite fast in the testing stage. Thus this learning-based method can detect the leakages of an FA image in a very short time. The rest of the paper is organized as follows: the datasets and methods are discussed in Sect. 2. The qualitative and quantitative comparison results are illustrated in Sect. 3. In Sect. 4, we summarize this work and discuss the experimental results.

2 Method

The flow chart of the proposed unsupervised leakage detecting method is illustrated in Fig. 1. We firstly train a CycleGAN-based network to generate the normal-looking FA images from the input abnormal images. Then, an abnormal FA image can be put into the trained network and generate a normal image. At last, the abnormal image subtracts the generated normal image, and a binarization processing is applied, the fluorescein leakages can be detected. Of course, the input image can be normal, then the network will output an image that is essentially unchanged from the input.

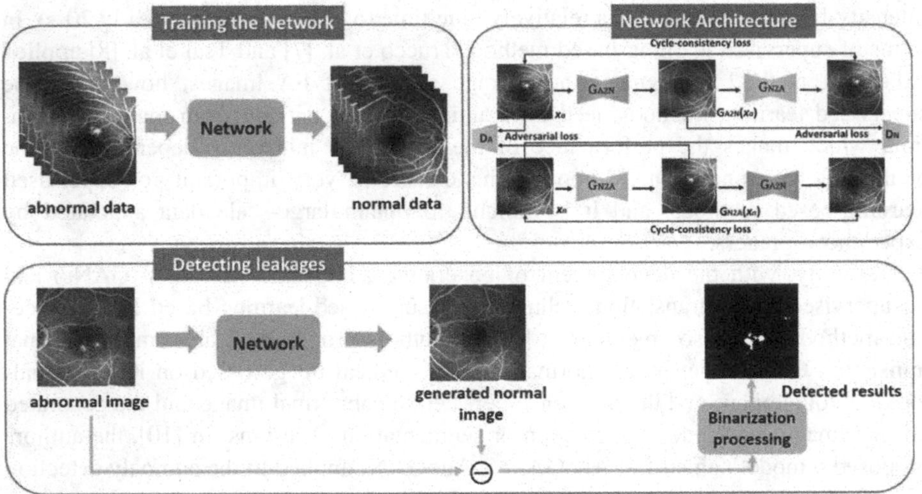

Fig. 1. The flow chart of the proposed unsupervised leakage detecting method. Firstly, a network that designed based on CycleGAN was trained to generate the normal-looking images from the input abnormal images. Then, the abnormal image can be put into the trained network to generate the normal-looking image. At last, making a difference between the generated normal and the abnormal image and doing binarization processing, the leakages can be detected.

2.1 Datasets

The datasets used in this work were taken with a Spectralis HRA (Heidelberg Engineering, Heidelberg, Germany) between March 2011 and September 2019 at the Third People's Hospital of Changzhou (Jiangsu, China), which called “HRA dataset” here.

The image types in our "HRA dataset" including normal FA images and the abnormal FA images with the optic disc, large focal, and punctate focal leakage (Fig. 2), which are the common types of leakage observed in most of the retinal diseases. The datasets initially contained 509 abnormal FA images and 343 normal FA images captured in the late stage of angiography from 852 eyes of 462 patients (223 female, 239 male, ranging in age from 7 to 86 years), and 1 picture per eye. The resolution of each image is 768 × 768 pixels, and the field of views of these images including 30°, 45°, and 60°. Twenty percent normal and abnormal FA images were randomly selected as the testing set, and the remainder data as the training set. Training data were augmented with random horizontal flips and rotations; moreover, we also use the FA image generation method [13] to generate FA images from the structure fundus images to further expand the scale of training data. Finally, after the data-augmentation operation, we totally have 2,062 abnormal and 1,970 normal FA images in the training set, respectively.

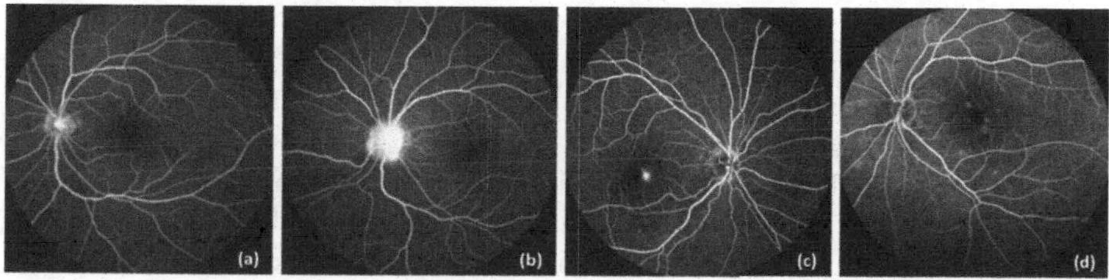

Fig. 2. Illustration of the types of FA image. (a) Normal FA image. FA image with (b) optic disc leakage; (c) large focal leakage; (d) punctate focal leakage.

2.2 Normal-Looking FA Image Synthesis and Leakage Detecting

Normal-Looking FA Image Synthesis

To transfer the abnormal FA image into the normal FA image domain, a CycleGAN model is utilized, which allows to perform domain transfers without the need for paired images. As shown in the upper right of Fig. 1, this network has two generators and two discriminators, which makes the network trained in an adversarial way. Generator G_{A2N} takes in an image x_a as input, and assumes to generate a normal-looking image $G_{A2N}(x_a)$, and the main goal of the generator G_{N2A} is to generate an image from the normal domain to abnormal domain. Two discriminators D_A and D_N aim to distinguish between real and generated images. The configuration of this network is similar to the original CycleGAN, which uses a ResNet with 9 residual blocks as the generators, and a 70 × 70 PatchGANs as the discriminators.

To make the network generate more realistic normal domain images, two loss functions are applied as [12] does, namely the adversarial loss L_{GAN}, and the cycle-consistency loss L_{cycle}, and the full loss function of this network can be written as:

$$\mathrm{L} = \lambda L_{GAN} + \mu L_{cycle}, \tag{1}$$

where λ and μ are the experimentally determined hyperparameters that control the effect of adversarial and cycle-consistency loss. In this task, we set $\lambda = 1$, $\mu = 10$, which were found to be the suitable parameters to achieve better performance.

Adversarial Loss: The proposed network adopts a bidirectional transform model with two generators G_{A2N} and G_{N2A} trained simultaneously. This strategy can help stabilize the model training via cycle consistency regularization. Since we have two generators and two discriminators, the GAN loss can be defined as:

$$L_{GAN} = \mathrm{E}_{p_a}[logD_A(x_a)] + \mathrm{E}_{p_a}[log(1 - D_N(G_{A2N}(x_a)))] \\ + \mathrm{E}_{p_n}[logD_N(x_n)] + \mathrm{E}_{p_n}[log(1 - D_A(G_{N2A}(x_n)))]. \quad (2)$$

Cycle Consistency Loss: The cycle consistency loss is adopted to transform normal and abnormal FA into one another and aid the learning of G_{A2N} and G_{N2A}.

$$L_{GAN} = \mathrm{E}_{p_a}\left[\|G_{N2A}(G_{A2N}(x_a)) - x_a\|_1\right] + \mathrm{E}_{p_n}\left[\|G_{A2N}(G_{N2A}(x_n)) - x_n\|_1\right]. \quad (3)$$

Angiographic Leakage Detecting

As illustrated in Fig. 1, after making the difference between abnormal and generated normal-looking image, binary processing is needed to obtain the leakage detecting results. In this work, we first use a 5 × 5 median filter to filter out the noise and eliminate the residual points that should not belong to the leakage areas. And then, the leakage detecting result is obtained by using the Otsu's Binarization method to binary the filtered image.

3 Experimental Results

To demonstrate the effectiveness of the proposed learning-based leakage detection method, we conducted a series of experiments on Zhao's dataset [2], Rabbani's dataset [5], and our HRA dataset. All experiments were carried out in an Ubuntu 16.04 + python 3.6 environment. We trained the proposed model for 200 epochs using Adam optimizer with initial learning rate of 0.0002, β_1 of 0.5 and β_2 of 0.999. After the first 100 epochs, the learning rate was decayed to zero over the next 100 epochs using cosine decay way. The proposed model was trained with a batch size of 2, based on the comprehensive consideration of the image generation performance and GPU memory consumption. It takes nearly 30 h on two GeForce GTX 1080Ti GPUs to train the model.

3.1 Evaluation Metrics for the Detecting Results

For a fair comparison with other state-of-the-art methods [2, 5], the testing results were all evaluated with the criteria of sensitivity (Sen), specificity (Spe), accuracy (Acc), area under the curve (AUC), and dice coefficient (DC), like that used in work [2] and [5]. And these metrics are defined as follows: $Sen = TP/(TP + FN)$; $Spe = TN/(TN + FP)$; $Acc = (TP + TN)/(TP + TN + FP + FN)$; $AUC = (Sen + Spe)/2$; $DC = 2(|A \cap B|)/(|A| + |B|)$, where *TP*, *TN*, *FP*, and *FN* indicate the true positive (correctly identified leakage pixels or regions), true negative (correctly identified background pixels or regions), false positive (incorrectly identified leakage pixels or regions), and false negative (incorrectly identified background pixels or regions), respectively. *A* represents the ground truth region, *B* is the segmented region, and $|A \cap B|$ denotes the number of pixels in the intersected region between *A* and *B*. All the pixels are equally treated towards their counting without considering the severity of the symptoms they depict.

3.2 Results on Open-Source Datasets

To demonstrate the effectiveness and universality of the proposed deep learning-based unsupervised method, the trained proposed model was tested on the open-source datasets that available at Zhao's and Rabbani's work [2, 5], and was compared quantitatively and qualitatively with their methods. Since we cannot get the source code of Zhao's and Rabbani's method, for these two public datasets, we directly quoted the results reported in their papers.

Results on Zhao's Datasets

Zhao's publicly available dataset contains 30 FA images and corresponding labeled images, with two types of leakage, large focal (20 images), and punctate focal leakage (10 images). All of these images had signs of malarial retinopathy (MR) on admission.

Figure 3 shows the leakage detection results over Zhao's dataset, it can be seen that most of the leakage areas were correctly identified by the proposed method. Table 1 illustrates the performances of different methods in detecting the focal leakage sites at the pixel level. It shows that the proposed method achieves comparable results to expert's annotation results, and it performs better than the other two state-of-the-art methods on most of the metrics. Although Zhao's method has good performance, the average specificity, accuracy, and dice coefficient of the proposed method all surpass Zhao's method 0.01, 0.01, and 0.02, respectively.

Noted that, since we do not have the source code of Zhao's method, we can only present two examples that include in both Zhao's publicly dataset and their paper. But we have segmented all the 20 focal images in their dataset using our method and gave the quantitative evaluation results. Thus, the evaluation results of the proposed method in Table 1 are the statistic results of a total of 20 image detecting results.

Results on Rabbani's Datasets

Rabbani's datasets contain 24 images captured from 24 subjects, and all the subjects had signs of diabetic retinopathy (DR) on admission. All the images in their datasets were categorized into three types according to the leakage conditions: predominantly focal (10 images), predominantly diffuse (7 images), and mixed pattern leakage (7 images).

The results of the proposed method over Rabbani's dataset are also compared quantitatively and qualitatively with Rabbani's and Zhao's method. As suggested in [5], quantitative analysis of a circular region centered at the fovea with a radius of 1500 μm is of the greatest significance for clinical diagnosis and treatment. For a fair comparative study, we also limited the proposed method of detecting the leakages in this area. As illustrated in Fig. 4, most of the leaking areas were all detected by the

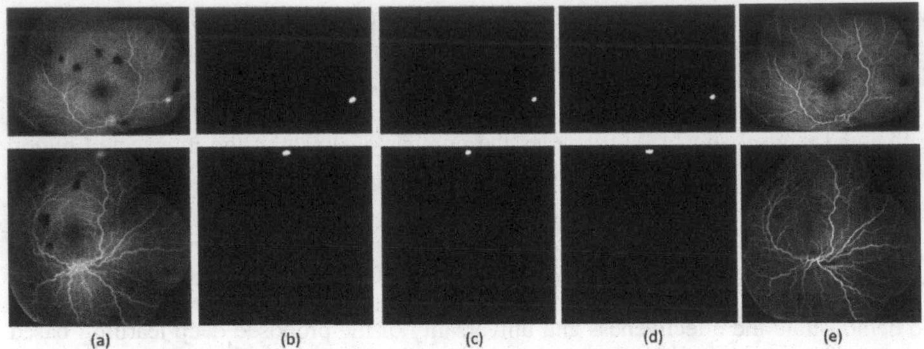

Fig. 3. Leakage detection results on Zhao's dataset. (a) Example FA image. Leakage segmented by: (b) expert's annotation; (c) Zhao's method; (d) the proposed method. (e) The normal-looking FA image generated by the proposed method.

Table 1. The performances of different methods on detecting the focal leakages over Zhao's Dataset at the pixel level. (* Data are expressed as average ± standard deviation)

	Intra obs.	Inter obs.	Rabbani's method	Zhao's method	Proposed method
Sen	0.96 ± 0.02	0.91 ± 0.04	0.81 ± 0.08	**0.93 ± 0.03**	0.92 ± 0.04
Spe	0.97 ± 0.03	0.94 ± 0.05	0.87 ± 0.08	0.96 ± 0.02	**0.97 ± 0.02**
Acc	0.96 ± 0.03	0.89 ± 0.04	0.83 ± 0.10	0.91 ± 0.03	**0.92 ± 0.02**
AUC	0.96 ± 0.02	0.92 ± 0.04	0.84 ± 0.08	**0.94 ± 0.02**	0.94 ± 0.03
DC	0.92 ± 0.04	0.80 ± 0.05	0.74 ± 0.05	0.82 ± 0.03	**0.84 ± 0.03**

three methods (Fig. 4(e)–(g)), and the segmentation results are very similar to the expert's annotation results. It is difficult to distinguish visually among the three methods, and the quantitative comparison is needed. Table 2 shows the statistic results of different methods of detecting the focal leakages over Rabbani's dataset at the pixel level. It can be seen that the proposed method outperforms the other two methods, whose average sensitivity, accuracy, and AUC is 0.02, 0.01, and 0.01 higher than Zhao's method, respectively.

3.3 Results on HRA Datasets

The proposed method was also tested on our HRA dataset. Since the FA images in our dataset do not have the manual annotation label by experts and we also do not have the source code of Zhao's and Rabbani's methods, we cannot compare the leakage detecting results with the label results and even Zhao's and Rabbani's results. Figure 5 shows the leakage detection results on our HRA dataset over the optic disc, large focal,

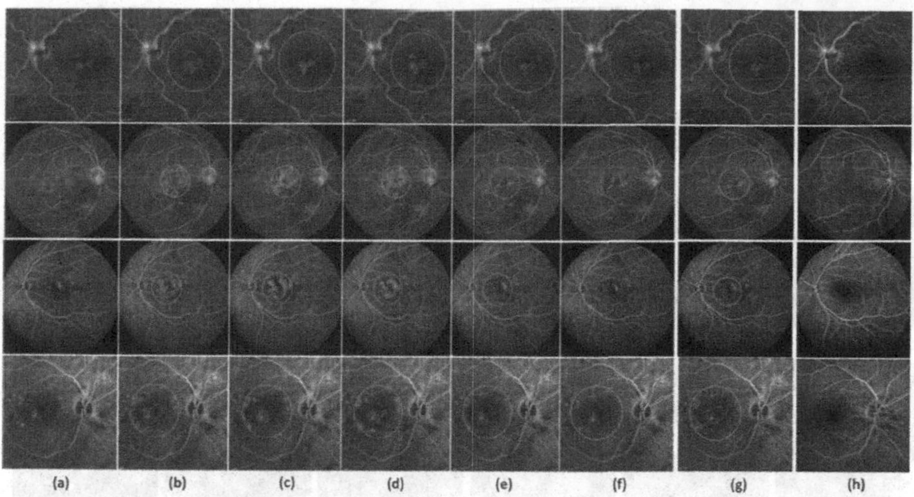

Fig. 4. Leakage detection results on Rabbani's dataset. (a) Example FA image. Leakage segmented by: (b) expert 1's annotation; (c) expert 2's annotation; (d) expert 2's annotation after 4 weeks; (e) Rabbani's method; (f) Zhao's method; (g) the proposed method. (h) The normal-looking FA image generated by the proposed method.

Table 2. The performances of different methods on detecting the focal leakages over Rabbani's Dataset at the pixel level. (* Data are expressed as average ± standard deviation)

	Intra obs.	Inter obs.	Rabbani's method	Zhao's method	Proposed method
Sen	0.95 ± 0.05	0.78 ± 0.09	0.69 ± 0.16	0.78 ± 0.06	**0.80 ± 0.06**
Spe	0.73 ± 0.27	0.94 ± 0.08	0.91 ± 0.09	**0.94 ± 0.02**	0.94 ± 0.04
Acc	0.83 ± 0.16	0.90 ± 0.08	0.86 ± 0.08	0.89 ± 0.06	**0.90 ± 0.05**
AUC	0.84 ± 0.16	0.91 ± 0.08	0.80 ± 0.12	0.86 ± 0.04	**0.87 ± 0.06**
DC	0.80 ± 0.08	0.82 ± 0.03	0.75 ± 0.05	**0.81 ± 0.02**	0.81 ± 0.04

and punctate focal leakages (Fig. 5 (a1)–(a6)), it can be seen that the proposed method can relatively accurately locate the leakages, even some small punctate leakages (Fig. 5 (c5) and (c6)). Moreover, as illustrated in Fig. 5(b1)–(b6), the proposed method can relatively accurately generate the normal-looking of the leaking areas, this is also a basis for the good leakage detection results.

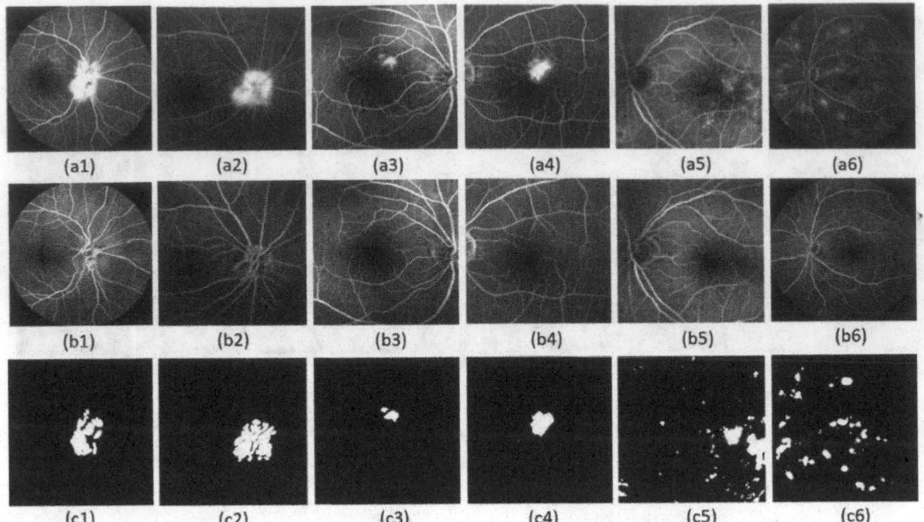

Fig. 5. Leakage detection results on HRA dataset over different types of leakage. Example FA images with: (a1)–(a2) optic disc leakage; (a3)–(a4) large focal leakage; (a5)–(a6) punctate focal leakage. (b1)–(b6) The normal-looking FA images generated by the proposed method; (c1)–(c6) Leakage segmentation results by the proposed method.

4 Discussion and Conclusion

Fundus fluorescein angiography (FA) can reflect the damaged state of the retinal barrier in eyes of living humans, especially in the early stage of diseases, and is regarded as a "good standard" of most retinal diseases diagnosis. Detecting the high-intensity retinal leakage in angiography is a key step for disease diagnosis and treatment. However, the manually labeled method is time-consuming and labor-intensive, thus an effective automated leakage detection method is warranted. At present, most automated leakage detection methods are an intensity-based method, this kind of method usually has relatively good performance, but it also has high time complexity, which needs nearly 20 s [2] to detect an image. In this study, we proposed a learning-based unsupervised leakage detecting method that can give the leakage detection results without any manual annotation and detect an image within 1 s.

The main idea of the proposed method is to train a model that can generate a corresponding normal-looking image from the input abnormal image; and then the leakages are detected by making the difference between the abnormal and the generated normal image. The generation of normal-looking images is the key step to accurately detect the leakages. We want the generator can automatically isolate and modify the leakage areas within the image while leaving any no-leakage areas within the image unchanged. However, it is difficult to achieve such an ideal state in practice and the proposed method also cannot, and this will lead false detecting of the leakages (The failure detection examples can be seen in Fig. 6 of Appendix part). Thus this should be the main problem to be solved in our future study. Although the proposed method performs not so perfect in leakage detection, this work demonstrates the effectiveness of the unsupervised learning-based method in fluorescein leakage detecting tasks.

To conclude, the proposed method has better performance when compared with other state-of-the-art methods whether in leakage detecting accuracy or time cost in detecting, which has great potential value for clinical diagnosis.

Appendix: Failure Detection Examples of the Proposed Method

See Fig. 6.

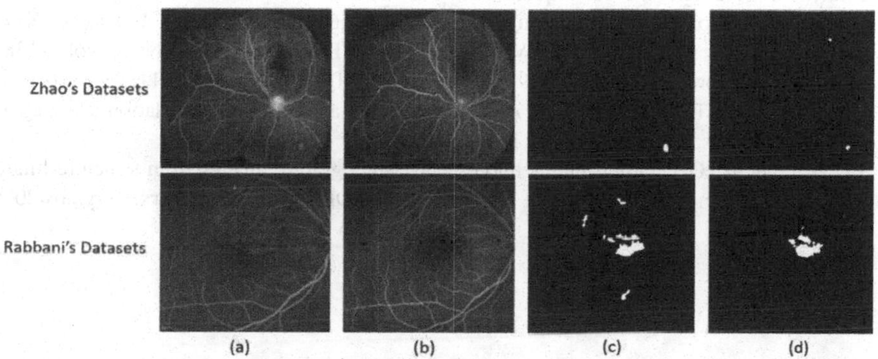

Fig. 6. Since the proposed method cannot only focus on the generation of leakage areas while leaving no-leakage area unchanged, it results in false detecting of the leakages. (a) Example abnormal FA image (b) The normal-looking FA image generated by the proposed method; (c) Expert's annotation; (d) Leakage detection results by the proposed method. (***Because of the limited space, we can only present one example from each dataset here***.)

References

1. Richard, G., Soubrane. G., Yannuzzi. L.A.: Fluorescein and ICG angiography: textbook and atlas. Thieme (1998)
2. Zhao, Y., et al.: Intensity and compactness enabled saliency estimation for leakage detection in diabetic and malarial retinopathy. IEEE Trans. Med. Imaging **36**, 51–63 (2016)
3. O'Toole, L.: Fluorescein and ICG angiograms : still a gold standard. Acta Ophthalmol. **85** (2007)
4. MartíNez-Costa, L., Marco, P., Ayala, G., Ves, E.De., Domingo, J., Simó, A.: Macular edema computer-aided evaluation in ocular vein occlusions. Comput. Biomed. Res. **31**, 374–384 (1998)
5. Rabbani, H., Allingham, M.J., Mettu, P.S., Cousins, S.W., Farsiu, S.: Fully automatic segmentation of fluorescein leakage in subjects with diabetic macular edema. 34–39 (2015). https://doi.org/10.1167/iovs.14-15457
6. Zhao, Y., et al.: Automated detection of leakage in fluorescein angiography images with application to malarial retinopathy. Sci. Rep. **5**, 10425 (2015)

7. Trucco, E., Buchanan, C.R., Aslam, T., Dhillon, B.: Contextual detection of ischemic regions in ultra-wide-field-of-view retinal fluorescein angiograms. In: Proceedings of the Annual International Conference of the IEEE Engineering in Medicine and Biology Society 2007, pp. 6740–6743 (2007)
8. Tsai, C., Ying, Y., Lin, W.: Automatic characterization of classic choroidal neovascularization by using AdaBoost for supervised learning automatic characterization and segmentation of classic choroidal neovascularization using Adaboost for supervised learning (2011). https://doi.org/10.1167/iovs.10-6048
9. Sun, L., Wang, J., Huang, Y., Ding, X., Greenspan, H., Paisley, J.: An adversarial learning approach to medical image synthesis for lesion detection. In: CVPR (2018)
10. Schlegl, T., Seeböck, P., Waldstein, S.M., Langs, G., Schmidt-erfurth, U.: f-AnoGAN: fast unsupervised anomaly detection with generative adversarial networks. Med. Image Anal. **54**, 30–44 (2019). https://doi.org/10.1016/j.media.2019.01.010
11. Baur, C., Wiestler, B., Albarqouni, S., Navab, N.: Deep autoencoding models for unsupervised anomaly segmentation in brain MR images. In: Crimi, A., Bakas, S., Kuijf, H., Keyvan, F., Reyes, M., van Walsum, T. (eds.) BrainLes 2018. LNCS, vol. 11383, pp. 161–169. Springer, Cham (2019). https://doi.org/10.1007/978-3-030-11723-8_16
12. Zhu, J.Y., Park, T., Isola, P., Efros, A.A.: Unpaired image-to-image translation using cycle-consistent adversarial networks. In: ICCV (2017)
13. Li, W.Y., et al.: Generating fundus fluorescence angiography images from structure fundus images using generative adversarial networks. In: MIDL (2020). https://arxiv.org/abs/2006.10216

Towards Automatic Embryo Staging in 3D+t Microscopy Images Using Convolutional Neural Networks and PointNets

Manuel Traub[1,2] and Johannes Stegmaier[1](✉)

[1] Institute of Imaging and Computer Vision, RWTH Aachen University, Aachen, Germany
johannes.stegmaier@lfb.rwth-aachen.de

[2] Institute for Automation and Applied Informatics, Karlsruhe Institute of Technology, Karlsruhe, Germany

Abstract. Automatic analyses and comparisons of different stages of embryonic development largely depend on a highly accurate spatiotemporal alignment of the investigated data sets. In this contribution, we assess multiple approaches for automatic staging of developing embryos that were imaged with time-resolved 3D light-sheet microscopy. The methods comprise image-based convolutional neural networks as well as an approach based on the PointNet architecture that directly operates on 3D point clouds of detected cell nuclei centroids. The experiments with four wild-type zebrafish embryos render both approaches suitable for automatic staging with average deviations of 21–34 min. Moreover, a proof-of-concept evaluation based on simulated 3D+t point cloud data sets shows that average deviations of less than 7 min are possible.

Keywords: Convolutional neural networks · PointNet · Regression · Transfer learning · Developmental biology · Simulating embryogenesis

1 Introduction

Embryonic development is characterized by a multitude of synchronized events, cell shape changes and large-scale tissue rearrangements that are crucial steps in the successful formation of a new organism [10]. To be able to compare these developmental events among different wild-type individuals, different mutants or upon exposure to certain chemicals, it is highly important to temporally synchronize acquired data sets, such that corresponding developmental stages are compared to one another [3,6]. While the temporal synchronization is typically easy

We thank the Helmholtz Association in the program BIFTM (MT), the German Research Foundation DFG (JS, Grant No STE2802/1-1) and the colleagues M. Takamiya, A. Kobitski, G. U. Nienhaus, U. Strähle and R. Mikut at the Karlsruhe Institute of Technology for providing the microscopy data and for the collaboration on previous analyses that form the basis of the data analyzed in this work.

N. Burgos et al. (Eds.): SASHIMI 2020, LNCS 12417, pp. 153–163, 2020.
https://doi.org/10.1007/978-3-030-59520-3_16

in small 2D screens, it becomes already a tedious undertaking when analyzing high-throughput screens consisting of thousands of repeats. Shifting dimensions to 3D the challenge of reproducible temporal synchronization becomes even more difficult and finally almost impossible for a human observer if the time domain is additionally considered. Current approaches to embryo staging largely rely on human intervention with the risk of subjective bias and might require specific labeling strategies or sophisticated visualization tools to cope with large-scale time-resolved 3D data [13,16].

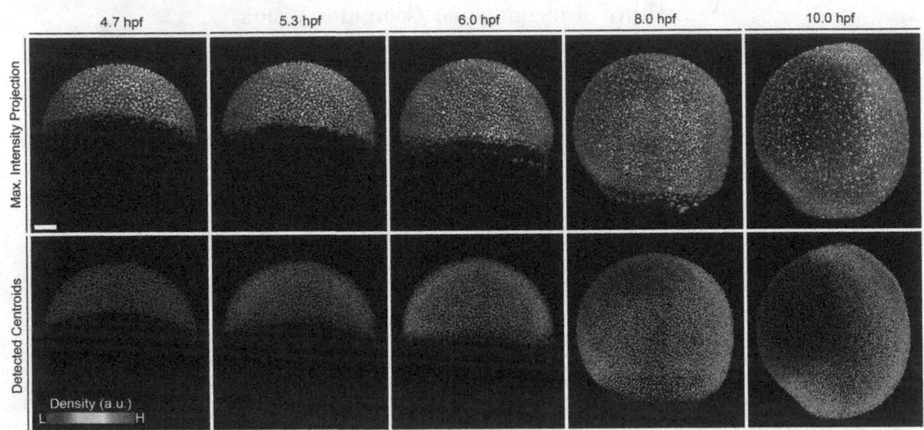

Fig. 1. Maximum intensity projections of 3D light-sheet microscopy images of a zebrafish embryo imaged from 4.7–10 hpf that were used as the input for the CNN-based stage predictions (top). Centroids of the fluorescently labeled cell nuclei were automatically detected [20] and are visualized with a density-based color-code that measures the relative number of neighbors in a fixed sphere neighborhood (bottom). A fixed number of points sampled from the centroids of different time points directly serves as input to the PointNet-based stage prediction. Scale bar: 100 µm.

Under the assumption that development progresses equally in all embryos, a rough temporal synchronization of the data sets can be obtained by measuring the time between fertilization and image acquisition [17]. Moreover, a visual staging can be performed after image acquisition by identifying certain developmental characteristics of a single snapshot of a time series or after chemical fixation of the embryo and by comparing it with a series of standardized views that show the characteristic development of a wild-type specimen at a standardized temperature [8]. In early phases of development, synchronized cell divisions regularly double the cell count and can be used to reliably align early time points [4,9]. Villoutreix *et al.* use measured cell counts over time to specify an affine transformation on the temporal domain, *i.e.*, the time axis is scaled such that the cell count curves of multiple embryos best overlap [22]. While cell counts might be reliable in specimens like the nemathode *C. elegans*, where adult individuals exhibit a largely identical number of cells, using the cell counts for synchronization in more complex animal models becomes more and more ambiguous. As development progresses, strong variation of total cell counts, cell sizes, cell shape and tissue formation are observed even among wild-type embryos [5,11]. If the number of individuals is limited and if later time points are considered, a manual identification of characteristic anatomical landmarks can be used to assign a specific developmental stage to selected frames [11,12,23]. With steadily increasing degrees of automation of experimental setups, it will be impractical to perform the staging with the human in the loop and automated approaches could thus help to further automate experimental protocols.

In this contribution, we analyzed two learning-based approaches for their suitability to automate embryo staging in large-scale 3D+t microscopy experiments. The methods comprise image-based convolutional neural networks (CNNs) as well as a point cloud-based approach using the PointNet architecture [14]. Both approaches were adapted for regression tasks and assessed under different hyperparameter and training conditions. We applied the methods to 2D maximum intensity projections as well as to real and simulated 3D point clouds of cell nuclei centroids (Fig. 1). Training data was extracted from terabyte-scale 3D+t light-sheet microscopy images of four wild-type zebrafish embryos (*D. rerio*) that ubiquitously expressed a green fluorescent protein (GFP) in their cell nuclei [9].

2 Automatic Embryo Staging as a Regression Problem

We analyzed two conceptually different approaches to automatically predict the hours post fertilization (hpf), a common measure for staging zebrafish embryos, either from 2D maximum intensity projection images or from 3D point clouds of centroids of fluorescently labeled cell nuclei. To estimate the achievable staging accuracy to be expected for training with a perfect ground truth, we created a synthetic 3D+t point cloud data set by adapting the method described in [19]. We intentionally did not use cell counts for staging due to large inter-specimen variations. Instead, the networks should learn to derive the stage solely from the appearance in the projection (CNN-based pipelines) or the arrangement of nuclei in space (PointNet-based pipeline).

2.1 CNN-Based Embryo Staging on Downsampled Maximum Intensity Projections

For the image-based approach we selected three established CNN architectures, namely VGG-16, ResNet-18 and GoogLeNet [7,18,21]. The classification output layers were replaced with a single regression node with a linear activation to predict the stage of the current input in hpf. In addition to the relatively large pretrained networks, we added a more shallow VGG-like network (VGG-Simple) consisting of four blocks of convolutions followed by ReLU activation (3×3 kernels, stride 1, 32 layers in the first convolutional layer and doubling the depth after each pooling operation), three max-pooling layers (2×2 kernels, stride 2) and two fully-connected layers (128 and 64 nodes, each followed by a dropout layer with probability $p = 0.2$) that map to the final regression node (Fig. 2).

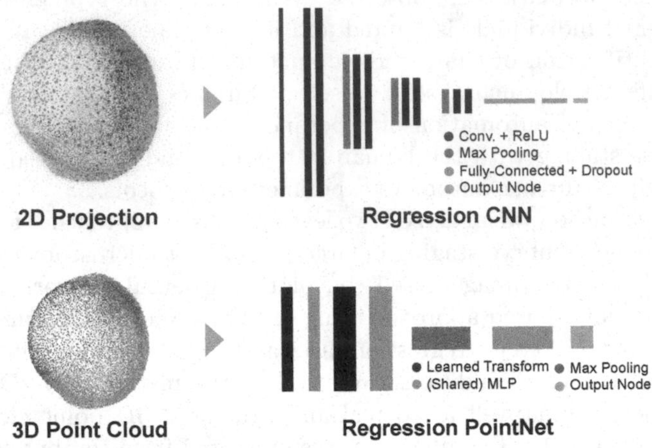

Fig. 2. Regression approaches used for automatic embryo staging. Different CNNs are applied to 2D maximum intensity projection images (top) and a PointNet [14] is applied to 3D point clouds of detected cell nucleus centroids (bottom). The depicted regression CNN architecture (top) is referred to as VGG-Simple in the remainder of the paper.

2.2 Regression PointNet for Automatic Embryo Staging

To perform the automatic staging on 3D point clouds, we equipped the original PointNet architecture by Qi *et al.* [14] with a regression layer as the target instead of classification scores (Fig. 2, bottom). In brief, the PointNet architecture processes input points separately by first putting them through learnable input and feature transformation modules (3×3 and 64×64) that are connected via a point-wise multilayer perceptrons with shared weights. This step is intended to orient the point cloud consistently and thus to obtain invariance with respect to input transformations. The transformed points are then fed through another set of multilayer perceptrons with shared weights (64, 128 and 1024 nodes in the hidden layers), a max-pooling operation that serves as a symmetric function to compensate for permutations of the input points and a final fully-connected multilayer perceptron with 512 and 256 hidden layers that map to k output scores in the original classification PointNet. To use PointNet for regression, we change the final fully-connected layer to map to a single output neuron with linear activation and use a mean squared error loss to regress to the target measure in hpf. As the PointNet operates only on a subset of the input points during each forward pass, we use an ensemble average of the predictions over 25 runs with randomly selected input points for the final output.

2.3 Synthetic Embryo Data Set to Identify the Possible Accuracy

As the temporal comparability of real embryos may be biased by the manual synchronization or due to inter-experiment variability caused by deviations from the assumption of the linear development, we decided to add an additional validation experiment using synthetic 3D+t point clouds that mimic embryonic development. As described in [19], virtual agents can be placed on tracked cell centroids of a real embryo, even in the presence incomplete or erroneous tracks. However, we simplified the approach described in [19], while still visually improving the realism of the simulations. The major difference is that the simulation is performed backwards in time, *i.e.*, we start at the last time point and initialize the simulation with a fraction of $p \in [0,1]$ randomly selected real positions at this time point. Each sampled position $\mathbf{x}_i$ was additionally modified by adding a small displacement in a random direction at most half the distance to its spatially nearest neighbor. So even if the same real cells were sampled, the additional randomization of each location ensures that no identical positions are present when creating multiple simulations of a single embryo. We skip the repulsive and nearest neighbor attraction forces and only rely on the movement directions of the closest neighbors in the real data set to update the positions of the simulated objects. The total displacement vector $\Delta\mathbf{x}_i^{\text{tot}}$ for each simulated object i at location $\mathbf{x}_i$ thus simplifies to the average of the backward directions $-\mathbf{d}_j$ of its K spatially nearest neighbors $j \in \mathcal{N}_{\text{knn}}^{K}(\mathbf{x}_i)$ within the real embryo:

$$\Delta\mathbf{x}_i^{\text{tot}} = \Delta\mathbf{x}_i^{\text{dir}} = \frac{1}{K} \sum_{j \in \mathcal{N}_{\text{knn}}^{K}(\mathbf{x}_i)} -\mathbf{d}_j. \tag{1}$$

As we start with the maximum number of objects and perform a time-reversed simulation, we need to incorporate merge events rather than cell division events. The number of merges N_k^{merge} required in time point k is defined as the difference of the objects currently present in the simulation N_k^{sim} and the target number of objects determined by the desired fraction of the real objects $p \cdot N_k^{\text{embryo}}$:

$$N_k^{\text{merge}} = \max\left(0, N_k^{\text{sim}} - p \cdot N_k^{\text{embryo}}\right). \tag{2}$$

To determine which of the current objects should be merged to preserve the density distribution of the real embryo even in the case of fewer simulated objects, we adapted the relative density difference measure from [19] and compute it for each object i at time point k:

$$\rho_{ik}^{\text{diff}} = \frac{\rho_{ik}^{\text{sim}}}{N_k^{\text{sim}}} - \frac{\rho_{ik}^{\text{embryo}}}{N_k^{\text{embryo}}}. \tag{3}$$

The density ρ_{ik} was defined as the number of neighbors present in a sphere with a fixed radius of 50 µm centered at each object in the real or the simulated point cloud, respectively. Although the simulation is not sensitive to the exact value selected for the neighborhood radius, it should be large enough that even in sparse regions of the simulation at least a few neighbors are present. Moreover, it is important to use the same radius for both the real and the simulated embryo, to ensure comparable density differences after normalizing the number of neighbors with the number of total objects at each time point. We select the N_k^{merge} simulated cells at time point k with the maximum relative density difference ρ_{ik}^{diff} and combine each of these objects with its spatially closest neighbor at time point k to a single fused object at time point $k-1$. As the PointNet-based stage identification is most relevant for practical applications, we did not simulate artificial image data and only focused on a 3D+t point cloud-based scenario for the identification of the achievable accuracy for a perfect ground truth.

3 Experiments

3.1 Data Acquisition and Training Data Generation

Raw image data was acquired using a custom-made light-sheet microscope as described in [9]. Experiments considered in this contribution comprise four wild-type zebrafish embryos expressing a transgenic H2A-GFP fusion protein in their cell nuclei that were imaged from about 2.5 to 16 hpf and were performed in accordance with the German animal protection regulations, approved by the Regierungspräsidium Karlsruhe, Germany (Az. 35-9185.64) [9]. A single experiment acquired with this technique accumulates about 10 terabytes of raw image data, *i.e.*, the total amount of data analyzed was on the order of 40 TB. Cell nuclei of all four embryos were automatically segmented using the TWANG segmentation algorithm [2,20] and we aligned the 3D+t point clouds in a consistent orientation with the prospective dorsal side pointing to the positive x-axis, the anteroposterior axis oriented along the y-axis [9,16]. A temporal window of 4.7–10 hpf was selected by manually identifying the 10 hpf stages as a synchronization time point and assuming constant development prior to that stage. The time interval was selected, such that the segmentation worked reliably with sufficiently strong fluorescent signal in the early time points and the ability to still resolve single cell nuclei in high density areas in later frames [16]. Stage identification was performed on 3D+t point clouds visualized in KitWare's ParaView [1]. The manual stage identification required several hours of interactive data visualization and analysis, which could largely benefit from automation.

For the image-based CNNs, we computed 2D maximum intensity projections for all data sets and all time points along the axial direction. Intensity values of all frames were scaled to the 8 bit range and contrast adjusted such that 0.3% of the pixels at the lower and the upper end of the intensity range were saturated. For networks pretrained on the ImageNet database, we scaled the single channel 8 bit images to 224×224 pixels with three redundant channels. We ended up with 370 frames for each embryo spanning a duration of 4.7–10 hpf, *i.e.*, 1480 images for all embryos in total. The PointNet was trained on 3D centroids of detected nuclei, which were extracted from the same four embryos as for the image-based experiments (between 4160 to 19794 per data set) [16]. In addition to the real data sets, we created four simulations using $p = 0.75$, *i.e.*,

75% of the real detections and movements of a single embryo but with different random initializations, such that we obtained the same overall shape and the same density distribution over time. The target hpf values were identical to the ones we used for the real embryo spanning from 4.7–10 hpf over 370 frames. The achievable performance is only assessed using the regression PointNet, as this is the practically most relevant approach and also allows to assess the 3D shape of the embryo irrespective of its orientation in the sample chamber rather than a 2D projection with potential occlusions.

3.2 Implementation and Training Details

The CNN-based approaches and pretrained networks were implemented in MATLAB using the Deep Learning Toolbox. We used the ADAM optimizer, tried different mini-batch sizes of 8, 16, 32 and 64 and trained for 100 epochs (sufficient for all models to converge). Optionally, we used data augmentation including reflection, scaling (0.9–1.0), random rotations (0–180°) and random translations (±5 px). In addition to training the networks from scratch, we investigated the performance of fixing the pretrained weights of the networks at different layers and only fine-tuned the deeper layers to the new task. For VGG-16, we tested all positions prior to the max-pooling layers (5 possibilities), for ResNet-18, all positions before and after pairs of ResNet modules (5 possibilities) and for GoogLeNet, all locations between Inception modules (9 possibilities). The PointNet was implemented using TensorFlow by modifying the original repository [14]. Training was performed using the ADAM optimizer and we randomly sampled 4096 centroids of each time point at each training iteration as the fixed-size input to the PointNet. We consistently obtained better results when disabling dropout and batch normalization during training and thus only report these results. For augmentation, random rotations around the origin were applied to all points and point jittering was used to apply small random displacements to each of the 3D input points. Training was performed with 4-fold cross-validation using three embryos for training and one embryo for testing in each fold, respectively. Simulated embryos were created using MATLAB and processed like the real data. Reported average values and standard deviations were computed on all folds.

4 Results and Discussion

The results of the best combination of architecture, augmentation and hyperparameters are summarized in Table 1 and Fig. 3. The smallest mean deviation from the ground truth was obtained with the PointNet approach with an average deviation of 0.35 ± 0.24 h (mean ± std. dev.). ResNet-18 was the best scoring CNN-based approach with an average deviation of 0.45 ± 0.30 h with a mini-batch size of 32 and with freezing the pretrained weights of the first four ResNet modules. These results were closely followed by the GoogLeNet trained from scratch with enabled data augmentation and a mini-batch size of 8 (0.50 ± 0.37 h) and the VGG-Simple architecture (0.51 ± 0.40 h) with augmentation enabled and a mini-batch size of 16. VGG-16 yielded the poorest performance in all investigated settings and the best results reported in Table 1 (0.57 ± 0.41 h) were obtained using a fine-tuning approach with keeping the weights up to the third convolutional block fixed, with enabled data augmentation and a mini-batch size of 8. The 138 million parameters of VGG-16 combined with a limited amount of training images showed a tendency to overfit despite using data augmentation and dropout regularization. However, with fixed pretrained weights and enabled data augmentation the qualitative trend did not change and the network even failed to learn the training data properly, which is probably due to the differences of the ImageNet data from fluorescence microscopy images. The same ranking was obtained when considering the root mean square deviation (RMSD, Table 1). To approximate the potentially possible accuracy, we assessed the performance of the PointNet on simulated data, reaching an average deviation of 0.11 ± 0.09 h (6.6 ± 5.4 min). Both in the early as well as in the late frames the major differences originate from cell density rather than shape, which is not yet considered by the models. Thus, the best performance was consistently achieved in intermediate frames (≈50–300) which are characterized by large-scale tissue rearrangements during the gastrulation phase.

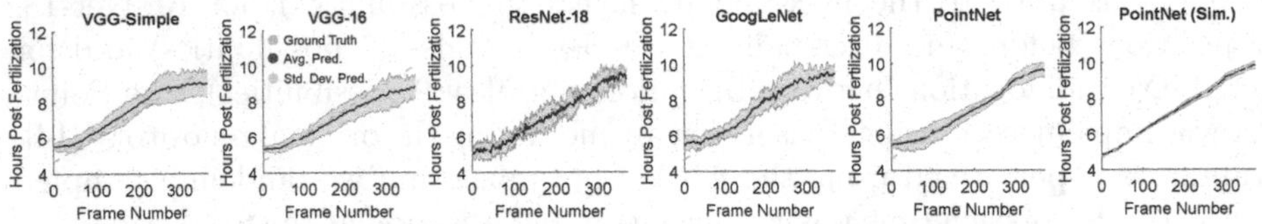

Fig. 3. Four-fold cross-validation results for the CNN-based predictions on maximum intensity projections (VGG-Simple, VGG-16, ResNet-18 and GoogLeNet, respectively) and the results of the regression PointNets (right). Ground truth is depicted in green and the mean ± std. dev. of the four cross-validation runs are depicted with black lines and gray shaded areas. Note that the PointNet (Sim.) approach was trained on synthetic data obtained from a single embryo to approximate the achievable accuracy. (Color figure online)

The performance of the image-based projection approach directly depends on the orientation of the embryo during image acquisition. On the contrary, the point cloud-based method is essentially independent of the initial orientation and randomized orientation of the point clouds are explicitly used as augmentation strategy during the training to further improve the invariance of the stage prediction with respect to the spatial orientation. The staging can thus be performed even without a tedious manual alignment of different experiments or the use of registration approaches, which might be required for the image-based approach. Robustness with respect to specimen orientation could be achieved for the image-based CNN approach by using 3D CNNs that are directly applied on downsampled raw image stacks [15]. However, this would require extensive preprocessing of the terabyte-scale data sets like multi-view fusion and downsampling, which would thus dramatically increase the required processing time. Currently, the time required for predicting the stage of a single time point is on the order of milliseconds for all investigated approaches.

Table 1. Staging accuracy of the different methods with top-scoring methods highlighted in bold-face letters. Last row: results obtained on the simulated data sets.

Method	Modality	Mean dev. ± Std. dev. (h)	RMSD (h)
VGG-Simple	2D Images	0.51 ± 0.40	0.65
VGG-16 [18]	2D Images	0.57 ± 0.41	0.70
ResNet-18 [7]	2D Images	0.45 ± 0.30	0.54
GoogLeNet [21]	2D Images	0.50 ± 0.37	0.62
PointNet [14]	3D Point Clouds	**0.35 ± 0.24**	**0.42**
PointNet [14]	3D Point Clouds (Sim.)	0.11 ± 0.09	0.14

The general average trends obtained by PointNet, ResNet-18 and GoogLeNet nicely resemble the ground truth and are arguably the most promising candidates among the presented methods. While an average deviation of about 21–30 min from the real developmental time can be readily used for a global staging and temporal alignment of long-term experiments, applications involving rapid tissue changes that potentially can happen in a matter of a few minutes might demand for a higher accuracy. However, as the experiments with simulated data shows, given sufficiently accurate ground truth data an accuracy of less than 7 min becomes possible. Moreover, we also expect a higher variability of the images in phases of rapid tissue changes that could potentially improve the separability. We consider the presented methods as a first proof-of-concept for automatic embryo staging in 3D+t experiments showcasing two conceptually different approaches. To compensate the lack of training data, we used several data augmentation strategies including various image transformations, point jittering and random rotations of the entire data set. However, data augmentation in the image-space did not consistently improve the results. As a straightforward extension, we could simply enlarge the training set by additional experiments or pre-train the real networks on realistic simulations as the ones presented in Sect. 2.3. Moreover, the data sets we used for demonstration were manually synchronized and assumed constant development of the embryo prior to the synchronization time point, with a potential subjective bias. To obtain a more objective ground truth calibration experiments would be required, *e.g.*, by imaging multiple channels of reporter genes that are known to be expressed within a distinct time window. Finally, an automatic spatial registration of the temporally aligned data sets will be the next logical step to eventually be able to automatically register multiple data sets in space and time.

References

1. Ayachit, U.: The Paraview Guide: A Parallel Visualization Application. Kitware Inc., New York (2015)
2. Bartschat, A., et al.: XPIWIT - an XML pipeline wrapper for the insight toolkit. Bioinformatics **32**(2), 315–317 (2016)
3. Castro-González, C., et al.: A digital framework to build, visualize and analyze a gene expression atlas with cellular resolution in zebrafish early embryogenesis. PLoS Comput. Biol. **10**(6), e1003670 (2014)
4. Faure, E., et al.: A workflow to process 3D+time microscopy images of developing organisms and reconstruct their cell lineage. Nat. Commun. **7**(8674), 1–10 (2016)
5. Fowlkes, C.C., et al.: A quantitative spatiotemporal atlas of gene expression in the drosophila blastoderm. Cell **133**(2), 364–374 (2008)
6. Guignard, L., et al.: Spatio-temporal registration of embryo images. In: Proceedings of the IEEE International Symposium on Biomedical Imaging, pp. 778–781 (2014)
7. He, K., Zhang, X., Ren, S., Sun, J.: Deep residual learning for image recognition. In: Proceedings of the IEEE Conference on Computer Vision and Pattern Recognition, pp. 770–778 (2016)
8. Kimmel, C.B., et al.: Stages of embryonic development of the Zebrafish. Dev. Dynam.: Off. Pub. Am. Assoc. Anatomists **203**(3), 253–310 (1995)
9. Kobitski, A., et al.: An ensemble-averaged, cell density-based digital model of zebrafish embryo development derived from light-sheet microscopy data with single-cell resolution. Sci. Rep. **5**(8601), 1–10 (2015)
10. Lecuit, T., Le Goff, L.: Orchestrating size and shape during morphogenesis. Nature **450**(7167), 189 (2007)
11. McDole, K., et al.: In toto imaging and reconstruction of post-implantation mouse development at the single-cell level. Cell **175**(3), 859–876 (2018)
12. Muenzing, S.E., et al.: larvalign: aligning gene expression patterns from the larval brain of drosophila melanogaster. Neuroinformatics **16**(1), 65–80 (2018)
13. Pietzsch, T., Saalfeld, S., Preibisch, S., Tomancak, P.: BigDataViewer: visualization and processing for large image data sets. Nat. Methods **12**(6), 481–483 (2015)
14. Qi, C.R., Su, H., Mo, K., Guibas, L.J.: PointNet: deep learning on point sets for 3D classification and segmentation. In: Proceedings of the IEEE Conference on Computer Vision and Pattern Recognition, pp. 652–660 (2017)
15. Qi, C.R., et al.: Volumetric and multi-view CNNs for object classification on 3D data. In: Proceedings of the IEEE Conference on Computer Vision and Pattern Recognition, pp. 5648–5656 (2016)
16. Schott, B., et al.: Embryominer: a new framework for interactive knowledge discovery in large-scale cell tracking data of developing embryos. PLoS Comput. Biol. **14**(4), 1–18 (2018)
17. Shahid, M., et al.: Zebrafish biosensor for toxicant induced muscle hyperactivity. Sci. Rep. **6**(23768), 1–14 (2016)
18. Simonyan, K., Zisserman, A.: Very deep convolutional networks for large-scale image recognition pp. 1–14. arXiv preprint arXiv:1409.1556, (2014)
19. Stegmaier, J., et al.: Generating semi-synthetic validation benchmarks for embryomics. In: Proceedings of the IEEE International Symposium on Biomedical Imaging, pp. 684–688 (2016)
20. Stegmaier, J., et al.: Automated prior knowledge-based quantification of neuronal patterns in the spinal cord of zebrafish. Bioinformatics **30**(5), 726–733 (2014)

21. Szegedy, C., et al.: Going deeper with convolutions. In: Proceedings of the IEEE Computer Society Conference On Computer Vision and Pattern Recognition, pp. 1–9 (2015)
22. Villoutreix, P., et al.: An integrated modelling framework from cells to organism based on a cohort of digital embryos. Sci. Rep. **6**, 1–11 (2016)
23. Winkley, K., Veeman, M.: A temperature-adjusted developmental timer for precise embryonic staging. Biol. Open **7**(6), bio032110 (2018)

Train Small, Generate Big: Synthesis of Colorectal Cancer Histology Images

Srijay Deshpande(✉), Fayyaz Minhas, and Nasir Rajpoot

Tissue Image Analytics Lab, University of Warwick, Coventry, UK
{srijay.deshpande,fayyaz.minhas,n.m.rajpoot}@warwick.ac.uk

Abstract. The construction of large tissue images is a challenging task in the field of generative modeling of histopathology images. Such synthetic images can be used for development and evaluation of various types of deep learning methods. However, memory and computational processing requirements limit the sizes of image constructed using neural generative models. To tackle this, we propose a conditional generative adversarial network framework that learns to generate and stitch small patches to construct large tissue image tiles while preserving global morphological characteristics. The key novelty of the proposed scheme is that it can be used to generate tiles larger than those used for training with high fidelity. Our evaluation of the Colorectal Adenocarcinoma Gland (CRAG) dataset shows that the proposed model can generate large tissue tiles that exhibit realistic morphological tissue features including glands appearance, nuclear structure, and stromal architecture. Our experimental results also show that the proposed model can be effectively used for evaluation of image segmentation models as well.

Keywords: Computational pathology · Generative adversarial networks · Image synthesis

1 Introduction

Automated analysis of histopathology whole slide images has gained immense attention in the field of digital pathology with applications such as segmentation [1], nuclei detection and classification [2–4] for facilitating clinical diagnosis. However, large image sizes of multi-gigapixel whole slide histopathology images pose the researchers with an unavoidable challenge of designing efficient networks to process these images. Along with it, there has been an increasing need to acquire data to evaluate image analytic tools. As a result, a wide variety of synthetic image generation algorithms have been developed for evaluation of such applications. For example, Kovacheva et al. [5] presented a tool to generate synthetic histology image data with parameters that allow control over the cancer grade and cellularity.

Recently, the field of computational pathology has witnessed a surge of deep learning methods over traditional machine learning techniques as they have

N. Burgos et al. (Eds.): SASHIMI 2020, LNCS 12417, pp. 164–173, 2020.
https://doi.org/10.1007/978-3-030-59520-3_17

shown significant improvement over the previous state-of-the-art results in the field of visual understanding and also in generative modeling. Generative Adversarial Networks (GAN) [6] have been popular in the recent years for image construction due to their ability to generate realistic images. Development of synthetic images conditioned on ground truth data through conditional GANs [7] is also an active area of research. Senaras et al. (2018) [8] have proposed a method to generate tissue images controlled by nuclei segmentation masks. Similarly, Quiros et al. (2019) proposed Pathology-GAN [9] based on generative adversarial networks to generate cancer tissue images. Such synthetic data can be potentially very useful in training and evaluation of machine learning models with limited real data. Though these models have been able to generate realistic tissue images successfully, existing methods are not well suited for generating large synthetic images or tiles due to their memory and processing capacity requirements. Large tiles usually provide a broader context in designing computational pathology algorithms and can be useful in performance assessment of such methods.

In this work, we propose a novel method for the generation of large synthetic histopathology tiles using a conditional generative adversarial network framework. The proposed framework can generate colorectal tissue images of arbitrary sizes by learning to generate and stitch smaller-sized patches, thus overcoming computational and memory size limitations. To the best of our knowledge, this is the first framework that can generate tiles of arbitrary sizes while maintaining global coherence. Key highlights of our proposed framework are:

1. It can generate tissue tiles of much larger sizes than ones used for training. This is beneficial as training models on such large images is computationally costly due to limitations in processing power and memory.
2. It shows the ability to learn to generate tissue regions, methodically stitch them to achieve homogeneous, realistic and globally coherent images without any seams.

In the next section, the details of the proposed framework are presented along with its mathematical formulation, followed by experiments on the CRAG dataset[1] [10]. The visual, qualitative, and quantitative performance assessment to evaluate our model is also provided.

2 Materials and Methods

Due to high computational complexity and memory limitations, neural image generation methods such as generative adversarial networks [6] and variational auto-encoders [11] typically restrict the sizes of generated images. In this paper, we propose a novel method for generating a large histology tissue tile from a given tissue component mask. The proposed scheme learns to generate small

[1] https://warwick.ac.uk/fac/sci/dcs/research/tia/data/mildnet.

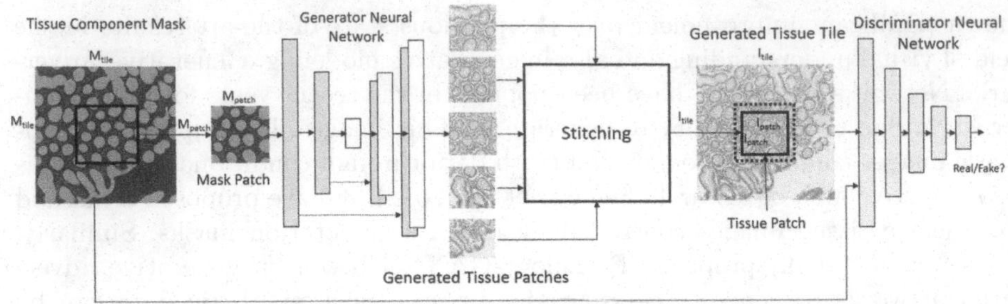

Fig. 1. Concept diagram of the proposed framework (Color figure online)

image patches based on local tissue components such as glands, stroma, background, etc. specified in the input mask and stitches the generated patches in a context-dependent and seamless manner thus overcoming computational limitations. Below, we discuss the details of the proposed method.

2.1 Dataset

For training a synthetic image generator for generating large images, we require annotated image data with large image sizes. Large datasets of such annotated data are typically not available. We have used the Colorectal Adenocarcinoma Gland Dataset (CRAG) for our experiments [10,12]. It contains 213 colon histopathology tissue images of large size of 1512 × 1516 pixels with variable glandular morphology and has been widely used for training and performance evaluation of histology image segmentation and glandular morphology analysis methods. Each image belongs to one of three classes based on its glandular morphology: normal (no deformation of glands), low grade (slightly deformed glands), and high grade (highly deformed glands). Each image in the dataset has a corresponding gland segmentation mask which specifies glandular segments in the image. We divide non-glandular regions from each mask into two parts, background and non-glandular tissue contents (mainly, stroma) giving the tissue component mask consisting of glands (shown in green), stroma (red), and background (blue) as shown in the leftmost part of Fig. 1. In this work, we have focused on generation of synthetic images with normal glands only by using the given segmentation mask as a tissue component mask. The dataset contains 48 large images with normal glands, out of which 39 (train set) are used to extract training data and 9 (test set) to extract testing data in this work.

2.2 The Proposed Framework

In order to generate large histology images from a given tissue component mask, we propose a conditional generative adversarial network based framework. The generative adversarial network (GAN) consists of two components: generator and discriminator. In a conventional GAN, the two networks are trained simultaneously such that the generator learns to produce realistic images from an underlying image distribution whereas the discriminator learns to discriminate between real images and the images generated by the generator. Similar to existing methods for histology image generation, the proposed approach also uses a generator network to generate small image patches from a given input tissue component mask. These image patches are then stitched to generate large image tiles. Unlike other existing methods, the proposed approach uses a discriminator network that works at the tile-level thus enforcing global coherence with minimal computational overhead. Since the patch-level generator is responsible for generating patches of the tissue tiles from corresponding input mask patches and the stitching of adjacent patches is independent of tile size, the framework can be used to construct large histology images of arbitrary size. Intuitively, the framework learns to generate local regions and stitch them in a way that, the generated tile exhibits seamless appearance and edge-crossing continuities between regions. The concept diagram of the proposed framework is shown in Fig. 1.

Patch-Level Generation. We use a set of training histology images and their associated tissue masks for training the proposed framework. For training, we extract square tiles of size $I_{tile} = 728$ with a stride of 200 from the train set of 39 large images, resulting in a total of 975 samples. Similarly for testing, we extract 144 square tiles of size $I_{tile} = 728$, 81 tiles of size $I_{tile} = 964$, 81 tiles of $I_{tile} = 1200$ and 36 tiles of size $I_{tile} = 1436$ from the test set of 9 large images (no overlap with training set). We denote a given histology image tile and its associated tissue component mask in our dataset as Y_i and X_i, respectively. Each tile in the given dataset can be modeled as a set of overlapping patches of a fixed size, i.e., $X = \{x_{r,c}\}$ and $Y = \{y_{r,c}\}$ with each patch parameterized by its center grid coordinates (r, c) in the corresponding tile. We use a generator network to generate an image patch y' given its corresponding input mask x, i.e., $y' = G(x; \theta_G)$, where θ_G represents the trainable weights of the generator. To train the generator in a context-aware manner, we present it with a larger $M_{patch} \times M_{patch}$ context patch of the input mask to generate a smaller patch of size $I_{patch} \times I_{patch}$ corresponding to the center of the input mask as shown in the Fig. 1 with $M_{patch} = 296$ and $I_{patch} = 256$ pixels.

Patch Stitching and Discriminator Modeling. While dividing the tissue component mask into patches, the stride $s = 236$ is chosen such that the generated patches are allowed to overlap by a small amount. This helps seamless stitching of the generated images to make a single tissue tile $Y' = \{y'_{r,c}\}$. The pixel values of the overlapping regions are spatially averaged in stitching.

The major challenge while generating the tissue tile, is to preserve global coherence and seamless appearance. For this purpose, the whole generated tile Y' is consumed by the discriminator network $D(X, Y'; \theta_D)$ with trainable weight parameters θ_D.

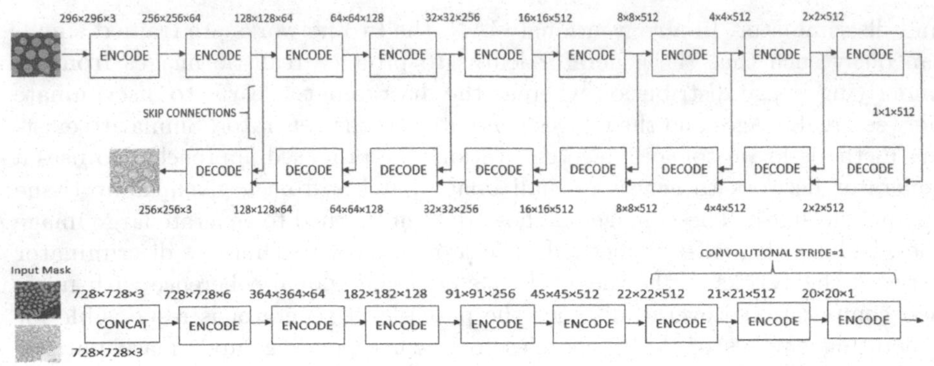

Fig. 2. Architectures of the generator (above) and the discriminator (below)

Generator and Discriminator Architecture. We adapt the architectures of the generator and discriminator architectures given in [13] for the proposed model. The generator is an encoder-decoder U-shaped structure similar to U-net [14]: it takes a tissue component patch as input, develops its low-dimensional representation through encoder blocks, and decompresses this representation via decoder blocks to generate a tissue image patch. The encoder is a series of "ENCODE" blocks where each block consists of a convolution layer followed by a batch normalization layer and a leaky-ReLU activation unit. The decoder is the chain of "DECODE" blocks with a deconvolution layer that up-samples the input to a higher dimension, followed by a batch normalization layer and leaky-ReLU activation unit. We use the version of the generator with skip-connections given in [13], which gives the flexibility to bypass the encoding part.

Unlike the proposed framework, the sizes of the input image and output image for the generator used in pix2pix conditional GAN network [13] are the same. In our experiment where the generator input size is 296×296 and output size is 256×256, we assemble an additional encoding block for the encoder part in the generator. The first encoder block reduces to the size of the image from 296 to 256 as shown in Fig. 2. After this step, the rest of the architecture is symmetric with skip connections connecting encoder layers to decoder layers.

We have used the PatchGAN discriminator [13] which accepts an input image mask and a tissue image and decides whether the second image is generated by generator or not. The architecture of the discriminator consists of a series of "ENCODE" blocks where the output is a 20×20 image in which each pixel value represents the output of the discriminator for the corresponding patch of the tile. The architecture of the discriminator is shown in Fig. 2.

Training and Inference. The two neural networks (generator G and discriminator D) are trained simultaneously to learn their trainable weight parameters $\{\theta_G, \theta_D\}$ from the given dataset. For this purpose, we use a loss function with two components:

1. Reconstruction Loss: In order to model regeneration errors, we use a tile level reconstruction loss based on the output of the generator model after stitching. Specifically, the reconstruction loss is the expected error between generated and actual tile level images in the training data as given below:

$$L_R = E_{X,Y} \|Y - G(X;\theta_G)\|_1 \tag{1}$$

2. Adversarial Loss: An adversarial loss (Eq. 2) is employed to make sure that the generated image is perceptually sound and looks realistic.

$$L_{adv} = E_{X,Y \sim p_{XY}}[logD(X,Y;\theta_D)] + E_{X \sim p_X}[log(1 - D(X, G(X,\theta_G);\theta_D)] \tag{2}$$

A weighted linear combination of the two losses is used as a joint optimization problem ($\min_{\theta_G} \max_{\theta_D} \lambda_R L_R + \lambda_{adv} L_{adv}$) with the weights as hyper-parameters which are tuned through cross-validation to 1.0 and 100.0, respectively. The network is trained with 100 epochs (Adam optimizer with an initial learning rate 10^{-4} and batch size 1). We update θ_G for each training patch whereas a single update in θ_D is performed for each tile.

It is important to note that the proposed framework is trained to generate globally consistent patches given an input tissue component mask. Consequently, at inference time, we can use the framework to generate a large image of arbitrary size as the stitching mechanism is independent of the tile size. We can also notice that this framework can be used to generate exhaustive annotated synthetic data using existing as well as methodically/randomly generated tissue component masks.

3 Results and Discussion

In order to validate the performance of the proposed framework on test data, we have conducted three types of assessments: 1. Visual Assessment, 2. Quantitative Assessment by using Frechet Inception Distance and Structural Similarity metrics, and, 3. Assessment Through Gland Segmentation in which we compared the concordance between gland segmentation accuracy of original and synthetic images.

3.1 Visual Assessment

Figure 3 shows the results of synthetic image generation from the proposed model, for different tile sizes. The first row shows tiles of size 1436×1436 whereas the second row shows tiles of size 964×964 pixels. It can be observed that the shapes of the glands are preserved in the constructed images. The constructed tiles appear seamless and homogeneous and show good preservation of morphological characteristics including epithelial cells, glandular regions, and stroma. Epithelial and goblet cells can be clearly distinguished in the glands with moderate deformities in glandular lumen.

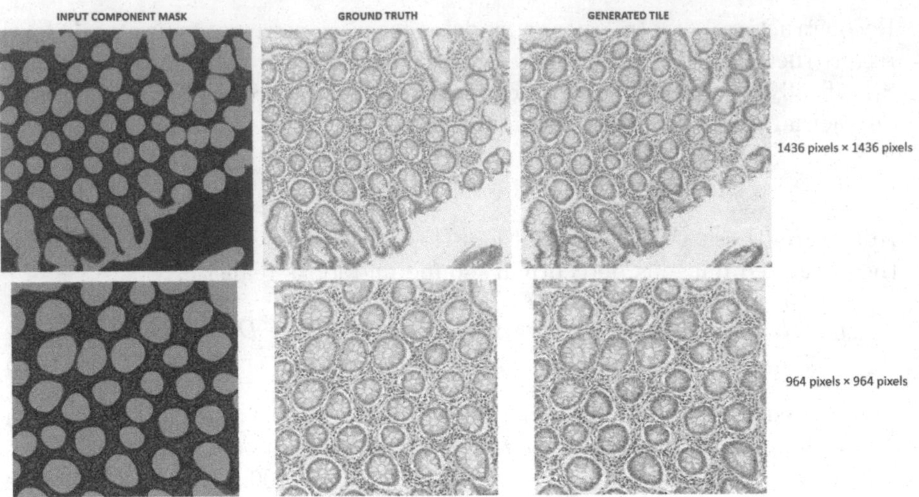

Fig. 3. Construction of large image tiles from input component masks

3.2 Quantitative Assessment

To evaluate the framework quantitatively we calculated the Frechet Inception Distance (FID) [15] and Structural Similarity Index (SSIM) [16] metrics which measures the degree of realism of the generated tiles.

Table 1. FID score comparison

Tile size	728 × 728		964 × 964		1200 × 1200		1436 × 1436	
Model	Random	Our model	Random	Our model	Random	Our model	Random	Our model
FID	130	**26**	74.5	**15.23**	74.75	**13.36**	57.82	**9.5**

Frechet Inception Distance. To compute FID [15], we collected features from the last pooling layer of the Inception V3 network [17] trained on the "ImageNet" dataset [18]. We generate the set of random noise images of the same dimensionality for the purpose of comparison. A low FID corresponds to high similarity between actual and generated tiles. Results in Table 1 show that the proposed algorithm generates significantly lower FID between real and synthetic tiles in comparison to random images. This indicates that convolution feature maps computed from original and synthetic images closely resemble each other in the feature space. This shows that the proposed framework is reliable for constructing synthetic data to be used for the experiments in the field of histopathology.

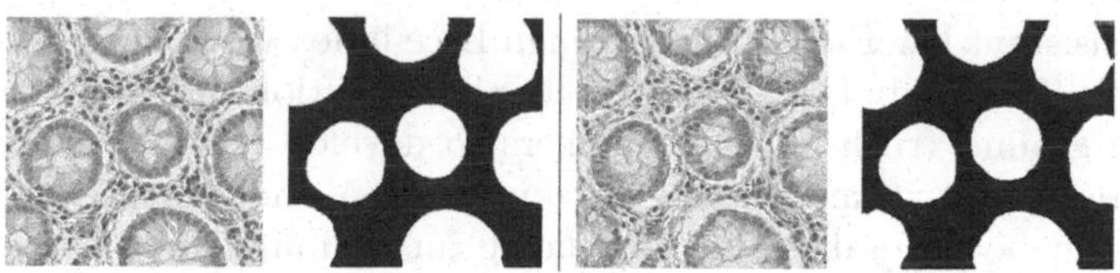

Fig. 4. Patches of original tiles (left) and generated tiles (right) along with their gland segmentation mask

Structural Similarity Index. We evaluated the Structural Similarity Index (SSIM) [16] to measure the similarity between the original and generated images. The score ranges between −1.0 & 1.0 where 1.0 denotes the highest structural similarity between images. The proposed method gives an average SSIM of 0.49 (with a standard deviation of 0.15) across the set of test images in our dataset. Though we can infer from the score that the generated tiles are similar to the original ones. However, the score is not very high because SSIM is dependent upon factors such as luminance and contrast as well, and the images in the data set have variable luminance and contrast among them.

3.3 Assessment Through Gland Segmentation

In this section, we show the performance of a U-net based gland segmentation algorithm [14] on synthetic data generated by our framework. For this purpose, 624 patches (of size 256 × 256) and 468 patches (of size 512 × 512) extracted from images of train set, are used to train the U-net segmentation model. Later, we computed binary gland segmentation masks on patches of both original and generated tiles, from images of test set (no overlap with train patches), and calculated the Dice index score among them (sample results shown in the Fig. 4).

We obtained an average Dice index [19] of 0.88 (with standard deviation 0.17) on 452 patches of 256 × 256 size, and a Dice index of 0.93 (with standard deviation 0.058) on 471 patches of 512 × 512 size, between segmentation masks of original and generated tiles. This high score conveys that it is reliable to use synthetic data constructed by our framework for evaluation of the gland segmentation algorithm.

4 Conclusions and Future Work

We have presented a novel framework for large histology image tile generation. We demonstrated that the framework can produce larger tiles even after being trained on relatively smaller tiles. The constructed tiles from the patch-based framework do not exhibit any seams or other deformities between adjacent patches. The results of the experiments performed on the CRAG dataset suggest that the generated tissue image tiles preserve morphological characteristics in the tissue regions. Additionally, the constructed images appear realistic and maintain consistent low FID scores and high Dice index when assessed via gland segmentation. The study has several practical applications as synthetic datasets with known ground truth allow researchers to develop state-of-the-art evaluations of various algorithms for nuclei segmentation and cancer grading. This framework can be generalized for producing an unlimited number of tiles for different types of carcinomas including low and high grades. This approach can be extended in the future to generate complete whole slide images from known parameters.

References

1. Qaiser, T., et al.: Fast and accurate tumor segmentation of histology images using persistent homology and deep convolutional features. Med. Image Anal. **55**, 1–14 (2019)
2. Graham, S., Rajpoot, N.M.: SAMS-NET: stain-aware multi-scale network for instance-based nuclei segmentation in histology images. In: 2018 IEEE 15th International Symposium on Biomedical Imaging (ISBI 2018), pp. 590–594. IEEE (2018)
3. Graham, S., et al.: HoVer-Net: simultaneous segmentation and classification of nuclei in multi-tissue histology images. Med. Image Anal. **58**, 101563 (2019)
4. Sirinukunwattana, K., Raza, S.E.A., Tsang, Y.W., Snead, D.R., Cree, I.A., Rajpoot, N.M.: Locality sensitive deep learning for detection and classification of nuclei in routine colon cancer histology images. IEEE Trans. Med. Imaging **35**(5), 1196–1206 (2016)
5. Kovacheva, V.N., Snead, D., Rajpoot, N.M.: A model of the spatial tumour heterogeneity in colorectal adenocarcinoma tissue. BMC Bioinform. **17**(1) (2016). Article number: 255. https://doi.org/10.1186/s12859-016-1126-2
6. Goodfellow, I., et al.: Generative adversarial nets. In: Advances in Neural Information Processing Systems, pp. 2672–2680 (2014)
7. Mirza, M., Osindero, S.: Conditional generative adversarial nets. arXiv preprint arXiv:1411.1784 (2014)
8. Senaras, C., et al.: Optimized generation of high-resolution phantom images using cGAN: application to quantification of Ki67 breast cancer images. PLoS ONE **13**(5), e0196846 (2018)
9. Quiros, A.C., Murray-Smith, R., Yuan, K.: Pathology GAN: learning deep representations of cancer tissue. arXiv preprint arXiv:1907.02644 (2019)
10. Graham, S., et al.: MILD-net: minimal information loss dilated network for gland instance segmentation in colon histology images. Med. Image Anal. **52**, 199–211 (2019)
11. Kingma, D.P., Welling, M.: An introduction to variational autoencoders. arXiv preprint arXiv:1906.02691 (2019)
12. Awan, R., et al.: Glandular morphometrics for objective grading of colorectal adenocarcinoma histology images. Sci. Rep. **7**(1), 1–12 (2017)
13. Isola, P., Zhu, J.Y., Zhou, T., Efros, A.A.: Image-to-image translation with conditional adversarial networks. In: Proceedings of the IEEE Conference on Computer Vision and Pattern Recognition, pp. 1125–1134 (2017)
14. Ronneberger, O., Fischer, P., Brox, T.: U-Net: convolutional networks for biomedical image segmentation. In: Navab, N., Hornegger, J., Wells, W.M., Frangi, A.F. (eds.) MICCAI 2015. LNCS, vol. 9351, pp. 234–241. Springer, Cham (2015). https://doi.org/10.1007/978-3-319-24574-4_28

15. Heusel, M., Ramsauer, H., Unterthiner, T., Nessler, B., Hochreiter, S.: GANs trained by a two time-scale update rule converge to a local nash equilibrium. In: Advances in Neural Information Processing Systems, pp. 6626–6637 (2017)
16. Wang, Z., Simoncelli, E.P., Bovik, A.C.: Multiscale structural similarity for image quality assessment. In: The Thrity-Seventh Asilomar Conference on Signals, Systems & Computers, vol. 2, pp. 1398–1402. IEEE (2003)
17. Szegedy, C., Vanhoucke, V., Ioffe, S., Shlens, J., Wojna, Z.: Rethinking the inception architecture for computer vision. In: Proceedings of the IEEE Conference on Computer Vision and Pattern Recognition, pp. 2818–2826 (2016)
18. Deng, J., Dong, W., Socher, R., Li, L.J., Li, K., Fei-Fei, L.: ImageNet: a large-scale hierarchical image database. In: 2009 IEEE Conference on Computer Vision and Pattern Recognition, pp. 248–255. IEEE (2009)
19. Zou, K.H., et al.: Statistical validation of image segmentation quality based on a spatial overlap index1: scientific reports. Acad. Radiol. **11**(2), 178–189 (2004)

Image Synthesis as a Pretext for Unsupervised Histopathological Diagnosis

Dejan Štepec[1,2](✉) and Danijel Skočaj[1]

[1] Faculty of Computer and Information Science, University of Ljubljana, Večna pot 113, 1000 Ljubljana, Slovenia
[2] XLAB d.o.o., Pot za Brdom 100, 1000 Ljubljana, Slovenia
dejan.stepec@xlab.si

Abstract. Anomaly detection in visual data refers to the problem of differentiating abnormal appearances from normal cases. Supervised approaches have been successfully applied to different domains, but require abundance of labeled data. Due to the nature of how anomalies occur and their underlying generating processes, it is hard to characterize and label them. Recent advances in deep generative based models have sparked interest towards applying such methods for unsupervised anomaly detection and have shown promising results in medical and industrial inspection domains. In this work we evaluate a crucial part of the unsupervised visual anomaly detection pipeline, that is needed for normal appearance modelling, as well as the ability to reconstruct closest looking normal and tumor samples. We adapt and evaluate different high-resolution state-of-the-art generative models from the face synthesis domain and demonstrate their superiority over currently used approaches on a challenging domain of digital pathology. Multifold improvement in image synthesis is demonstrated in terms of the quality and resolution of the generated images, validated also against the supervised model.

Keywords: Anomaly detection · Unsupervised · Deep-learning · Generative adversarial networks · Image synthesis · Digital pathology

1 Introduction

Anomaly detection represents an important process of determining instances that stand out from the rest of the data. Detecting such occurrences in different data modalities is widely applicable in different domains such as fraud detection, cyber-intrusion, industrial inspection and medical imaging [1]. Detecting anomalies in high-dimensional data (e.g. images) is a particularly challenging problem, that has recently seen a particular rise of interest, due to prevalence of deep-learning based methods, but their success has mostly relied on abundance of available labeled data.

N. Burgos et al. (Eds.): SASHIMI 2020, LNCS 12417, pp. 174–183, 2020.
https://doi.org/10.1007/978-3-030-59520-3_18

Anomalies generally occur rarely, in different shapes and forms and are thus extremely hard or even impossible to label. Supervised deep-learning approaches have seen great success, especially evident in the domains with well known characterization of the anomalies and abundance of labeled data. Obtaining such detailed labels to learn supervised models is a costly and in many cases also an impossible process, due to unknown set of all the disease biomarkers or product defects. In an unsupervised setting, only normal samples are available (e.g. healthy, defect-free), without any labels. Deep generative methods have been recently applied to the problem of unsupervised anomaly detection (UAD), by utilizing the abundance of unlabeled data and demonstrating promising results in medical and industrial inspection domains [2–4]. Deep generative methods, in a form of autoencoders [5] or GANs [6] are in a UAD setting used to capture normal appearance, in order to detect and segment deviations from that normal appearance, without the need for labeled data.

AnoGAN [7] represents the first method, where GANs are used for anomaly detection in medical domain. A rich generative model is constructed on healthy examples of optical coherence tomography (OCT) images of the retina and a methodology is presented for image mapping into the latent space. The induced latent vector is used to generate the closest example to the presented query image, in order to detect and segment the anomalies in an unsupervised fashion. AnoGAN [7] and the recently presented f-AnoGAN [2] improvement, utilize low resolution vanilla DCGAN [8] and Wasserstein GAN [9] architectures, for normal appearance modelling, with significantly lower anomaly detection performance in comparison with autoencoder-based approaches [10,11]. This does not coincide with superior image synthesis performance of the recent GAN-based methods. We argue, that adapting recent advancements in GAN-based unconditional image generation [12–14], currently utilized mostly for human face synthesis, should also greatly improve the performance of image synthesis in different medical imaging domains, as well GAN-based UAD methods.

In this work we focus on normal appearance modelling part of the UAD pipeline and evaluate the ability to synthesize realistically looking histology imagery. This presents a pretext for an important problem of metastases detection from digitized gigapixel histology imagery. This particular problem has been already addressed in a supervised setting [15], by relying on the limited amount of expertly lesion-level labeled data, as well as in a weakly-supervised setting [16], where only image-level labels were used. Extremely large histology imagery and highly variable appearance of the anomalies (i.e. cancerous regions) represent a unique challenge for existing UAD approaches. We investigate the use of the recently presented high resolution generative models from the human face synthesis domain [12–14], for normal appearance modelling in a UAD pipeline, which could consequently improve the performance and stability of the current state-of-the-art approaches [2,11].

We demonstrate this with significant improvements in the quality and increased resolution of the generated imagery in comparison with currently used approaches [8,9], which also represents a novel application of generative models to the digital pathology domain. We also investigate the effectiveness of current latent space mapping approaches, specifically their ability of closest looking normal histology sample reconstruction. We validate the quality of the synthesized imagery against the supervised model and demonstrate the importance of synthesizing high resolution histology imagery, resulting in an increased amount of contextual information present, crucial for distinguishing tumor samples from the normal ones.

2 Methodology

Unsupervised Visual Anomaly Detection. The capability to learn the distribution of the normal appearance represents one of the most important parts in a visual anomaly detection pipeline, presented in Fig. 1. This is achieved by learning deep generative models on normal samples only, as presented in Fig. 1a. The result of this process is the capability to generate realistically looking artificial normal samples, which cannot be distinguished from the real ones. To detect an anomaly, a query image is presented and the closest possible normal sample appearance is generated, which is used to threshold the difference, in order to detect and segment the anomalous region, as presented in Fig. 1b. This is possible due to learned manifold of normal appearance and its inability to reconstruct anomalous samples.

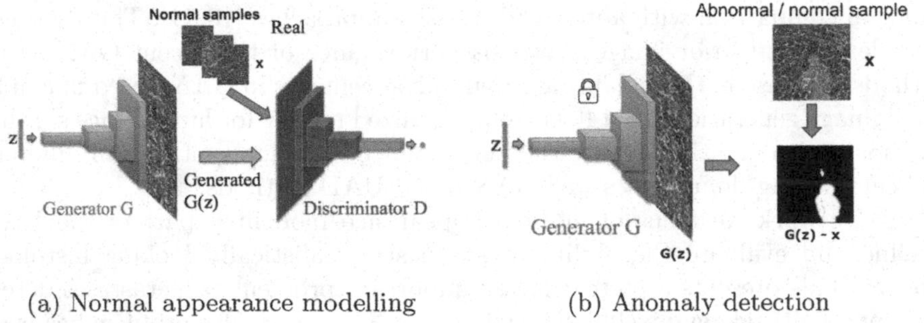

(a) Normal appearance modelling (b) Anomaly detection

Fig. 1. GAN based visual anomaly detection pipeline, consisting out of a) normal appearance modelling and b) the search for optimal latent representation, that will generate the closest normal appearance sample, used for anomaly detection.

Different approaches have been proposed for normal appearance modelling, as well as anomaly detection. Learning the normal visual appearance is based on autoencoders [3], GANs [2,7], or combined hybrid models [10,11]. Most of the approaches learn the space of the normal sample distribution Z, from which latent vectors $z \in Z$ are sampled from, that generate the closest normal appearance, to the presented query image. Different solutions have been proposed for latent vector optimization, that are usually independent from the used normal appearance modelling method (i.e. autoencoders, GANs).

Current state-of-the-art GAN-based visual anomaly detection methods [2,7] are based on vanilla GAN implementations [8,9], with limited resolution generated normal samples of low fidelity, as well as with stability problems. Autoencoder based anomaly detection methods have shown improved results over pure GAN based implementations [10,11], but are similarly limited to a low resolution of 64^2. In comparison, we evaluate the feasibility of generating high fidelity histology imagery up to the resolution of 512^2, with the recently presented GAN architectures [12–14] from the face generation domain, not yet utilized in anomaly detection pipelines, as well as in the digital pathology domain. Note, that we limit the resolution to a maximum of 512^2, due to hardware resource and time constraints. We also investigate the effectiveness of recently presented latent space mapping approaches and the feasibility to be applicable for UAD in the digital pathology domain.

Deep Generative Adversarial Models. Original GAN implementation [6] was based on standard neural networks, which generated images suffered from being noisy and the training process was notoriously unstable. This was improved by implementing the GAN idea using the CNNs - DCGAN [8], by identifying a family of architectures, that result in a stable training process of higher resolution deep generative models. The method was later on also adapted in the AnoGAN anomaly detection framework [7]. Stability problems of GAN methods were first improved by proposing different distance measures for the cost function (e.g. Wasserstein GAN [9]), adapted also by the f-AnoGAN anomaly detection method [2]. The main limitation of those early GAN methods is also the low resolution (up to 64^2) and the limited variability of the generated images.

Recently, the ideas of progressively growing GANs [12] and style-based generators [13,14] were presented, allowing a stable training of models for resolutions up to 1024^2, with increased variation and quality of the generated images. In progressively growing GANs [12], layers are added to the generator and discriminator progressively, by linearly fading them in and thus enabling fast and stable training. StyleGAN [13] proposes an alternative generator architecture, based on style transfer literature [17], exposing novel ways to control synthesis process and reducing the entanglement of the latent space. StyleGAN2 [14] addresses some of the main characteristic artifacts resulting from the progressive growing in StyleGAN [13], further boosting generative performance.

In comparison with autoencoders, GANs do not automatically yield the inverse mapping from the image to latent space, which is needed for closest-looking normal sample reconstruction and consequently anomaly detection. In AnoGAN [7] an iterative optimization approach was proposed to optimize the latent vector z via backpropagation, using the residual and discrimination loss. Residual loss is represented with pixel-wise Mean Square Error (MSE) loss, while discrimination loss is guided by the GAN discriminator, by computing feature matching loss between the real and synthesized imagery. In f-AnoGAN method [2], an autoencoder replaces the iterative optimization procedure, using the trainable encoder and the pre-trained generator, as the decoder. For Style-

GAN2 [14], authors proposed an iterative inverting procedure, which specifically optimizes an intermediate latent space and noise maps, based on the Learned Perceptual Image Patch Similarity (LPIPS) [18].

3 Experiments and Results

Histology Imagery Dataset. We address aforementioned problems of anomaly detection pipeline on a challenging domain of digital pathology, where whole-slide histology images (WSI) are used for diagnostic assessment of the spread of the cancer. This particular problem was already addressed in a supervised setting [15], as a competition[1], with provided clinical histology imagery and ground truth data. A training dataset with (n = 110) and without (n = 160) cancerous regions is provided, as well as a test set of 129 images (49 with and 80 without anomalies). Raw histology imagery, presented in Fig. 2a, is first preprocessed, in order to extract the tissue region (Fig. 2b). We used the approach from IBM[2], which utilizes a combination of morphological and color space filtering operations. Patches of different sizes (64^2–512^2) are then extracted from the filtered image (Fig. 2c) and labelled according to the amount of tissue in the extracted patch.

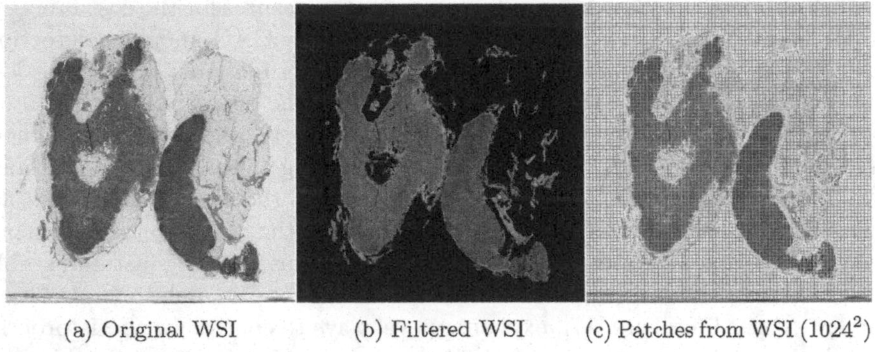

(a) Original WSI (b) Filtered WSI (c) Patches from WSI (1024^2)

Fig. 2. Preprocessing of the original WSI presented in a) consists of b) filtering tissue sections and c) extracting patches, based on tissue percentage (green ≥90%, red ≤10% and yellow in-between). Best viewed in digital version with zoom. (Color figure online)

Image Synthesis. We first evaluate the performance of different GAN based generative approaches on a challenging histology imagery using the Fréchet Inception Distance (FID) [19], by following evaluation procedure from StyleGAN2 [14, Table 1]. The FID score represents a similarity between a set of

[1] https://camelyon16.grand-challenge.org/.
[2] https://github.com/CODAIT/deep-histopath.

real and generated images, based on the statistics extracted from the pretrained Inception classifier [20]. We train different GANs using 1000 randomly extracted patches, from each of the 160 normal WSIs, with tissue coverage over 90% (Fig. 2c). This presents an input to the baseline DCGAN [8] model, used in the AnoGAN [7] anomaly detection framework, Wasserstein GAN (WGAN), used in f-AnoGAN [2], as well as to recently presented GAN architectures, based on progressive growing (PGAN [12]) and style transfer (StyleGAN [13], StyleGAN2 [14]). We evaluate not only the feasibility to generate pathology imagery, but to generate it up to a resolution of 512^2 and present the results in Table 1.

Table 1. FID scores for different methods and different input image sizes DCGAN and WGAN are limited to a maximum resolution of 64^2 and only best performing model is evaluated at 512^2.

Image size	DCGAN	WGAN	PGAN	StyleGAN	StyleGAN2
64^2	88.52	12.65	18.89	7.15	**6.64**
256^2	–	–	17.82	5.57	**5.24**
512^2	–	–	–	–	**2.93**

We can see that generative performance of the recently presented methods significantly outperforms vanilla DCGAN model and should be considered in all the proposed GAN based visual anomaly detection pipelines. They are also capable of generating images of much bigger size and higher resolution, which is particularly important for anomaly detection, due to the increased amount of visual context, available for determining the presence or absence of the anomalies. For WGAN, we used the implementation from f-AnoGAN [2], based on residual neural networks, which also produces high quality, high fidelity imagery up to image size of 64^2. There is no significant difference in terms of the FID score between different style transfer based approaches, but StyleGAN2 represents an incremental improvement over StyleGAN and also offers an additional benefit of generator reversibility, especially interesting for anomaly detection (Fig. 1b). Note that in StyleGAN2 a smaller network configuration E was used for an increased image throughput, at small performance expense [14, Table 1]. Visual comparison of generated results is presented in Fig. 3, demonstrating high variability and quality of the generated images and also the applicability of such methods for digital pathology.

Classification of Real and Synthetic Patches. To additionally asses the quality of the generated images in comparison with the real ones, we evaluated the performance of a discriminative classifier applied on both types of data. We trained supervised DenseNet121 [21] model on extracted normal and tumor histology imagery patches and compared its performance to distinguish the two classes on real and synthesized imagery (Table 2). We extracted 100.000 normal

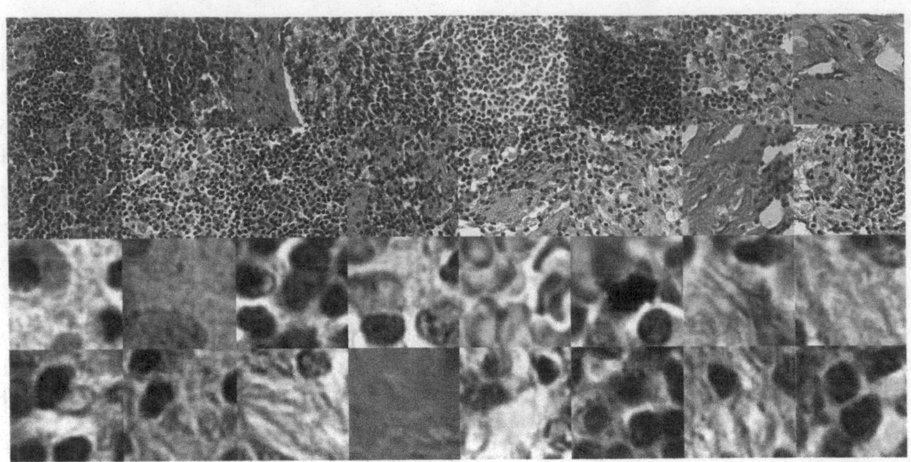

Fig. 3. Examples of real histology imagery at image size of 512^2 (top row), generated images by the best performing StyleGAN2 model at 512^2 (second row), real histology imagery at 64^2 (third row) and images generated by the WGAN model at 64^2 (last row). Best viewed in digital version with zoom. (Color figure online)

and tumor patches (with provided annotations) of size 64^2 and 512^2 to train DenseNet121 model and evaluated the performance on a test set of 10.000 (real) normal and tumor patches. Besides on normal histology imagery, we also trained WGAN [9] and StyleGAN2 [14] on tumor patches, in order to be able to generate both classes. We then synthesized a test set of 10.000 normal and tumor patches and evaluated the capability of the DenseNet121 model (trained on real imagery) to distinguish anomalous samples in synthesized imagery.

Table 2. Classification accuracy (CA) for normal vs. tumor patch based classification on real and synthesized imagery (WGAN for 64^2 and StyleGAN2 for 512^2).

Image size	Real (CA)	Synthesized (CA)
64^2	88.23%	87.05%
512^2	98.89%	98.34%

Results in Table 2 demonstrate that supervised model (trained on real imagery) successfully recognizes synthesized imagery, with no drop in performance, compared to real imagery. Supervised model can be seen as a virtual pathologist, confirming to some extent the correctness of image synthesis, on a much larger scale. We can also see significant drop in performance on 64^2 patches, which additionally confirms, that larger patches hold more contextual information, useful for classification.

Latent Space Mapping. We qualitatively evaluate the capability of latent space projection using the encoder based approach (izi_f) presented in f-AnoGAN [2] for WGAN [9] based generator and LPIPS-distance [18] based approach for StyleGAN2 [14] based generator, proposed and specifically designed for StyleGAN2 already in the original work [14]. Figure 4b presents the results for both methods, image sizes and histopathological classes (i.e. normal and tumor). The generators are trained on normal samples only and should reconstruct only normal samples, while tumor samples should be poorly reconstructed, thus enabling detection of anomalies. We can see that the encoder based approach, proposed in f-AnoGAN [2] is able to find very similar looking artificial samples in the latent space. The problem is, that it also reconstructs tumor samples, with similar performance. We argue, that this is due to small patch size (64^2), which is in more than 10% ambiguous also for the supervised classifier (Table 2) - such small tumor patch might in fact not contain abnormality, or is represented with insufficient biomarkers. Such samples, even in small percentage [22], cause the generator to learn how to reconstruct anomalous samples.

StyleGAN2 [14] did not yield good reconstructions, capturing only major properties of the query image. This is beneficial for tumor samples, where we noticed consistent failure to capture even the main properties of the query image, with notable exception of the staining color. This can probably be attributed to the larger image sizes and more contextual information present to distinguish anomalous samples, not seen during the generator training. Integration of StyleGAN2 and encoder based mapping (coupling best of the two methods) offers a promising future direction to be investigated, to improve latent space mapping and thus enabling UAD in histopathological analysis.

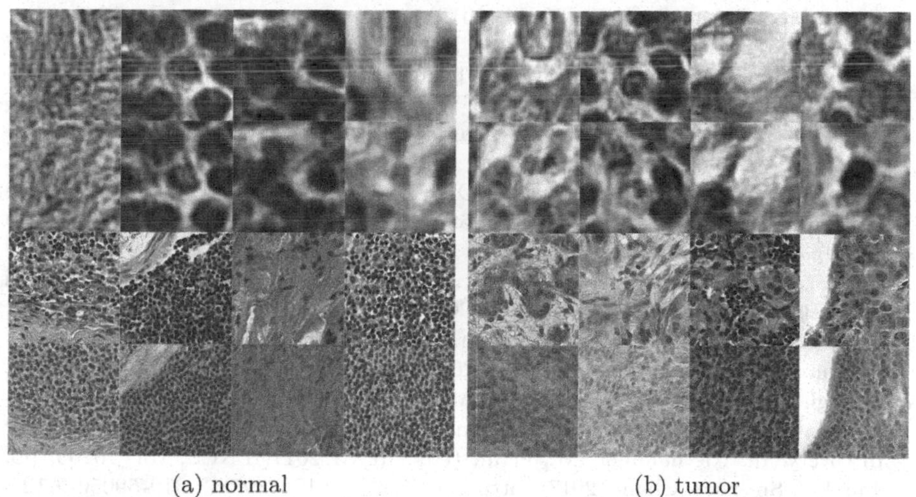

(a) normal (b) tumor

Fig. 4. Projecting real a) normal and b) tumor imagery to the latent space of WGAN at 64^2 (first row) and StyleGAN2 at 512^2 (third row) and resulting closest matches in the latent space for WGAN (second row) and StyleGAN2 (last row). Best viewed in digital version with zoom. (Color figure online)

4 Conclusion

In this work we addressed image synthesis as a pretext for GAN based anomaly detection pipeline in histopathological diagnosis and demonstrated, that histology imagery of high quality and variability can be synthesized, as well as reconstructed. We identified the importance of synthesizing large histology samples, not used in current GAN based anomaly detection pipelines, as well as the drawbacks and future research direction for more effective latent space mapping. The ability to generate realistically looking normal histology imagery of high resolution and size will enable the development of UAD pipeline, in order to apply it to cancer diagnosis, especially important for rare cancer types (e.g. paediatric), where annotated data is scarce, thus preventing the use of supervised approaches. Reducing the performance gap between supervised and unsupervised approaches and increasing the robustness of the UAD approaches will represent a significant contribution to wider adaption of automated visual analysis techniques, well beyond presented medical domain.

Acknowledgment. This work was partially supported by the European Commission through the Horizon 2020 research and innovation program under grant 826121 (iPC).

References

1. Chandola, V., Banerjee, A., Kumar, V.: Anomaly detection: a survey. ACM Comput. Surv. (CSUR) **41**(3), 1–58 (2009)
2. Schlegl, T., Seeböck, P., Waldstein, S.M., Langs, G., Schmidt-Erfurth, U.: f-AnoGAN: fast unsupervised anomaly detection with generative adversarial networks. Med. Image Anal. **54**, 30–44 (2019)
3. Baur, C., Wiestler, B., Albarqouni, S., Navab, N.: Deep autoencoding models for unsupervised anomaly segmentation in brain MR images. In: Crimi, A., Bakas, S., Kuijf, H., Keyvan, F., Reyes, M., van Walsum, T. (eds.) BrainLes 2018. LNCS, vol. 11383, pp. 161–169. Springer, Cham (2019). https://doi.org/10.1007/978-3-030-11723-8_16
4. Bergmann, P., Fauser, M., Sattlegger, D., Steger, C.: MVTec AD-a comprehensive real-world dataset for unsupervised anomaly detection. In: CVPR, pp. 9592–9600 (2019)
5. Xia, Y., Cao, X., Wen, F., Hua, G., Sun, J.: Learning discriminative reconstructions for unsupervised outlier removal. In: ICCV, pp. 1511–1519 (2015)
6. Goodfellow, I., et al.: Generative adversarial nets. In: Advances in Neural Information Processing Systems, pp. 2672–2680 (2014)
7. Schlegl, T., Seeböck, P., Waldstein, S.M., Schmidt-Erfurth, U., Langs, G.: Unsupervised anomaly detection with generative adversarial networks to guide marker discovery. In: Niethammer, M., et al. (eds.) IPMI 2017. LNCS, vol. 10265, pp. 146–157. Springer, Cham (2017). https://doi.org/10.1007/978-3-319-59050-9_12
8. Radford, A., Metz, L., Chintala, S.: Unsupervised representation learning with deep convolutional generative adversarial networks. In: ICLR (2016)
9. Martin Arjovsky, S., Bottou, L.: Wasserstein generative adversarial networks. In: ICML (2017)

10. Akcay, S., Atapour-Abarghouei, A., Breckon, T.P.: GANomaly: semi-supervised anomaly detection via adversarial training. In: Jawahar, C.V., Li, H., Mori, G., Schindler, K. (eds.) ACCV 2018. LNCS, vol. 11363, pp. 622–637. Springer, Cham (2019). https://doi.org/10.1007/978-3-030-20893-6_39
11. Akçay, S., Atapour-Abarghouei, A., Breckon, T.P.: Skip-GANomaly: skip connected and adversarially trained encoder-decoder anomaly detection. In: IJNN, pp. 1–8. IEEE (2019)
12. Karras, T., Aila, T., Laine, S., Lehtinen, J.: Progressive growing of GANs for improved quality, stability, and variation. In: ICLR (2018)
13. Karras, T., Laine, S., Aila, T.: A style-based generator architecture for generative adversarial networks. In: CVPR, pp. 4401–4410 (2019)
14. Karras, T., Laine, S., Aittala, M., Hellsten, J., Lehtinen, J., Aila, T.: Analyzing and improving the image quality of StyleGAN. In: CVPR, pp. 8110–8119 (2020)
15. Bejnordi, B.E., et al.: Diagnostic assessment of deep learning algorithms for detection of lymph node metastases in women with breast cancer. JAMA **318**(22), 2199–2210 (2017)
16. Campanella, G., et al.: Clinical-grade computational pathology using weakly supervised deep learning on whole slide images. Nat. Med. **25**(8), 1301–1309 (2019)
17. Huang, X., Belongie, S.: Arbitrary style transfer in real-time with adaptive instance normalization. In: CVPR, pp. 1501–1510 (2017)
18. Zhang, R., Isola, P., Efros, A.A., Shechtman, E., Wang, O.: The unreasonable effectiveness of deep features as a perceptual metric. In: CVPR, pp. 586–595 (2018)
19. Heusel, M., Ramsauer, H., Unterthiner, T., Nessler, B., Hochreiter, S.: GANs trained by a two time-scale update rule converge to a local nash equilibrium. In: Advances in Neural Information Processing Systems, pp. 6626–6637 (2017)
20. Szegedy, C., Vanhoucke, V., Ioffe, S., Shlens, J., Wojna, Z.: Rethinking the inception architecture for computer vision. In: CVPR, pp. 2818–2826 (2016)
21. Huang, G., Liu, Z., Van Der Maaten, L., Weinberger, K.Q.: Densely connected convolutional networks. In: CVPR, pp. 4700–4708 (2017)
22. Berg, A., Ahlberg, J., Felsberg, M.: Unsupervised learning of anomaly detection from contaminated image data using simultaneous encoder training. arXiv preprint arXiv:1905.11034 (2019)

Auditory Nerve Fiber Health Estimation Using Patient Specific Cochlear Implant Stimulation Models

Ziteng Liu(✉), Ahmet Cakir, and Jack H. Noble

Department of Electrical Engineering and Computer Science,
Vanderbilt University, Nashville, TN 37235, USA
ziteng.liu@vanderbilt.edu

Abstract. Cochlear implants (CIs) restore hearing using an array of electrodes implanted in the cochlea to directly stimulate auditory nerve fibers (ANFs). Hearing outcomes with CIs are dependent on the health of the ANFs. In this research, we developed an approach to estimate the health of ANFs using patient-customized, image-based computational models of CI stimulation. Our stimulation models build on a previous model-based solution to estimate the intra-cochlear electric field (EF) created by the CI. Herein, we propose to use the estimated EF to drive ANF models representing 75 nerve bundles along the length of the cochlea. We propose a method to detect the neural health of the ANF models by optimizing neural health parameters to minimize the sum of squared differences between simulated and the physiological measurements available via patients' CIs. The resulting health parameters provide an estimate of the health of ANF bundles. Experiments with 8 subjects show promising model prediction accuracy, with excellent agreement between neural stimulation responses that are clinically measured and those that are predicted by our parameter optimized models. These results suggest our modeling approach may provide an accurate estimation of ANF health for CI users.

Keywords: Cochlear implant · Auditory nerve fibers · Optimization

1 Introduction

Cochlear implants (CIs) are considered the standard-of-care treatment for profound sensory-based hearing loss. In normal hearing, sound waves induce pressure oscillations in the cochlear fluids, which in turn initiate a traveling wave of displacement along the basilar membrane (BM). This membrane divides the cochlea along its length and produces maximal response to sounds at different frequencies [1]. Because motion of BM is then sensed by hair cells which are attached to the BM, these sensory cells are fine-tuned to respond to different frequencies of the received sounds. The hair cells further pass signals to auditory nerve fibers (ANFs) by releasing chemical transmitters. Finally, the electrical stimulation is propagated along the ANFs to the auditory cortex allowing the brain to sense and process the sounds.

For patients suffering sensorineural hearing loss, which is principally caused by damage or destruction of the hair cells, direct stimulation of the auditory nerve using a

N. Burgos et al. (Eds.): SASHIMI 2020, LNCS 12417, pp. 184–194, 2020.
https://doi.org/10.1007/978-3-030-59520-3_19

CI is possible if ANFs are intact [2]. A CI replaces the hair cells with an externally worn signal processor that decomposes the incoming sound into signals sent to an electrode array that is surgically implanted into the cochlea (see Fig. 1a). Electrode arrays have up to 22 contacts depending on the manufacturer, dividing the available ANFs to, at most, 22 frequency bands or stimulation areas when using monopolar stimulation. Studies have shown that hearing outcomes with CIs are dependent on several factors including how healthy the ANFs are [14]. After surgery, CI recipients undergo many programming sessions with an audiologist who adjusts the settings for every single electrode to improve overall hearing performance. However, lacking objective information about ANF health and more generally about what settings will lead to better performance, a trial and error procedure is implemented. As weeks of experience with given settings are needed to indicate long-term outcome with those settings, this process can be frustratingly long and lead to suboptimal outcomes.

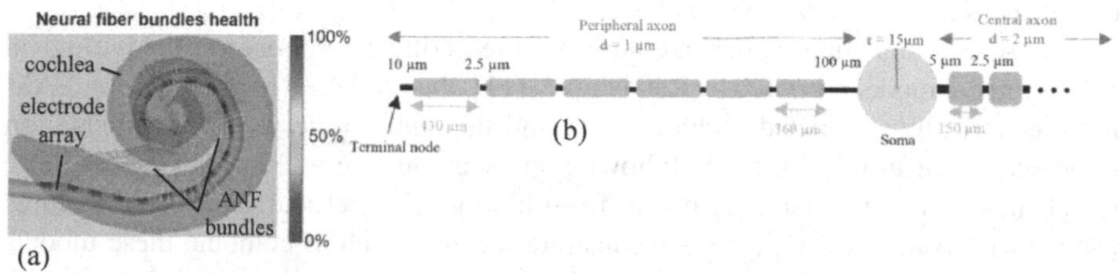

Fig. 1. Overview of the ANF models. (a) shows the spatial distribution of ANF bundles colored with a nerve health estimate. (b) Shows the ANF stimulation model created for each fiber bundle.

Our group has been developing image-guided CI programming techniques (IGCIP) in order to provide objective information that can assist audiologists with programming [3–5]. Although IGCIP has led to better hearing outcomes in experiments [4, 5], neural stimulation patterns of the electrodes are estimated in a coarse manner using only the distance from each electrode to the neural activation sites in our current implementation. So, it is possible that the method could be improved with a better estimate of the electrodes' neural activation patterns with a physics-based model. To achieve that, we developed patient-specific models of the electrically stimulated cochlea [6, 7] which allow us to estimate intra-cochlear electric fields (EF) created by the CI for individual patients. Building on those studies, in this study we propose to use these EF models as input to ANF activation models to predict neural activation caused by electrical stimulation with the CI. We also propose the first *in vivo* approach to estimate the health of individual ANFs for CI patients using these models.

In summary, herein we propose patient-customized, image-based computational models of ANF stimulation. We also present a validation study in which we verify the model accuracy by comparing its predictions to clinical neural response measurements. Our methods provide patient-specific estimation of the electro-neural interface in unprecedented detail and could enable novel programming strategies that significantly improve hearing outcomes with CIs.

2 Related Works

Several groups have proposed methods for predicting neural activation caused by electrical stimulation [11, 16, 17]. Most of these methods use physiologically-based active membrane nerve models driven by physics-based estimation of the voltage distribution within a given anatomical structure. However, these studies either lack the capacity to be applied in-vivo or only confine themselves to anatomical customization instead of constructing both anatomically and electrically customized models that take advantage of physiological measurements that are clinically available. It is possible that these models need to be fully customized in order to prove useful for clinical use. Thus, in this work we are proposing patient-customized, computational ANF stimulation models, which are not only coupled with our patient-specific electro-anatomical models (EAMs) to ensure electrical and anatomy customization, but also estimate neural health status along the length of cochlea. Our models permit accurately simulating physiological measurements available via CIs.

Our ANF stimulation models are built on three critical components: the biological auditory nerve model proposed by Rattay *et al.* [11], the CT-based high-resolution EAM of the electrically stimulated cochlea [6, 7], and the auditory nerve fiber segmentation proposed by our group [9]. In the following subsections, we will introduce how these models help to describe auditory nerves from biological, electrical, and spatial features respectively. And in Sect. 3, we will illustrate our approach to combine these models and build our novel, health-dependent ANF stimulation models based on them.

2.1 Biological Nerve Model

The model proposed in by Rattay *et al.* [11] introduce three major features that differs from other nerve models. First, they use compartment model which consists of several subunits with individual geometric and electric parameters as shown in Fig. 1b. Second, Ion channel dynamics are described by a modified Hodgkin-Huxley (HH) model, namely, 'warmed' HH (wHH) model. wHH includes sodium, potassium and leakage currents and has the following form:

$$\tfrac{dV}{dt} = [-g_{Na}m^3h(V - V_{Na}) - g_K n^4(V - V_K) - g_L(V - V_L) + i_{stimulus}]/c \tag{1}$$

$$\tfrac{dm}{dt} = [-(\alpha_m + \beta_m)m + \alpha_m]k \tag{2}$$

$$\tfrac{dh}{dt} = [-(\alpha_h + \beta_h)h + \alpha_h]k \tag{3}$$

$$\tfrac{dn}{dt} = [-(\alpha_n + \beta_n)n + \alpha_n]k \tag{4}$$

$$k = 3^{T-6.3} \tag{5}$$

$$V = V_i - V_e - V_{rest} \tag{6}$$

where V, V_i, V_e and V_{rest} are the membrane, internal, external and resting voltages, and V_{Na}, V_K, and V_L are the sodium, potassium and leakage battery voltages, respectively. g_{Na}, g_K, g_L are the maximum conductance and m, h, n are probabilities with which the maximum conductance is reduced with respect to measured gating data, for sodium, potassium, and leakage, respectively. $i_{stimulus}$ is the current produced by electrode stimulation, and c is the membrane capacity. Finally, α and β are voltage dependent variables that were fitted from measured data, k is the temperature coefficient, and T is temperature in Celsius. With wHH, the gating processes are accelerated (m, h, n are multiplied by 12), which best fit to observed temporal behavior of human auditory nerves compared to the original HH model, and leakage conductances are multiplied by the factor 10 to simulate 10-fold channel density. Also, the influence of membrane noise is also taken into account in their approach. These features allow the model to simulate the electrically excited auditory nerves in the human cochlea more accurately than models based on animals.

2.2 Electro-Anatomical Model and Auditory Nerve Fiber Segmentation

In a series of previous studies [6, 7], our group created CT-based high-resolution EAMs to determine the patient-specific EF caused by the current injected via CI electrodes. Briefly, this EAM estimates a volumetric map of the EF through the cochlea created by the CI. The EAM is customized for each patient by customizing a conductivity map so that estimated impedances between all combinations of the CI electrodes best match clinical measurements of these quantities (termed Electrical Field Imaging (EFI)). Then the EF can be found by solving Poisson's equation for electrostatics, which is given by $\nabla \cdot J = -\sigma \nabla^2 \Phi$, where Φ is the EF, J is the electric current density and σ is the conductivity. We are able to define the current source and ground for the CI versus other nodes by manipulating the left-hand side of the equation. As it is discussed in [6], the tissue in this model was assumed to be purely resistive, thus the amount of current enters a node equals to the amount of current that leaves the same node. The finite difference method solution to it can be foud by solving $A\vec{\Phi} = \vec{b}$, where A is a sparse matrix containing coefficients of the linear sum of currents equations, $\vec{\Phi}$ are the set of node voltages that are being determined and are concatenated into a vector, and $b(i)$ equals to $+1\,\mu\text{A}$ if the ith node is a current source and 0 otherwise. The nodes representing ground are eliminated from the system of linear equations, so the net current is not constrained for those nodes. This system of linear equations is then solved by using the bi-conjugate gradient method [6].

The EAMs are electrically customized by optimizing the tissue resistivity estimates to minimize the average error between simulated EFIs and measured EFIs. The resistivity values for different tissue classes, including electrolytic fluid, soft tissues, neural tissue, and bone, are bound to vary in a range of 50 to 150% of their default values, which are 300, 50, 600, and 5000 Ωcm respectively. Figure 2 shows the EFI simulation of a customized EAM and a generic EAM which uses default electrical properties for 4 electrodes of the same subject, demonstrating much better agreement between simulated and measured EFI after customizing electrical properties.

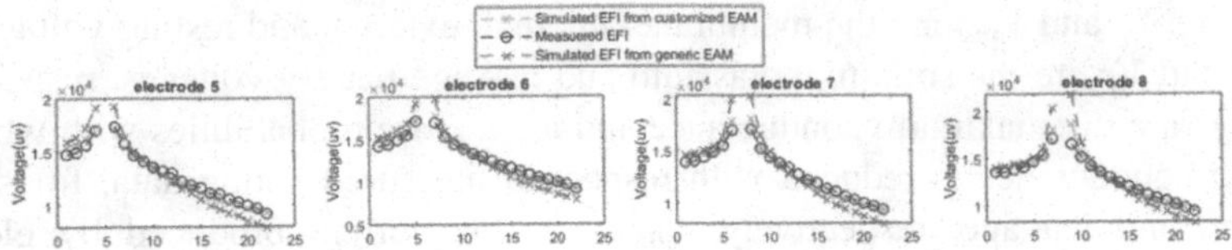

Fig. 2. EFI simulation of a customized EAM and a generic EAM

To localize the ANFs, we use a semi-automatic segmentation technique proposed in [8]. That approach relies on prior knowledge of the morphology of the fibers to estimate their position. It treats the fiber localization problem as a path-finding problem [8]. Several points are automatically defined as landmarks using the segmentation of the cochlea. Paths representing 75 fiber bundles that are evenly spaced along the length of the cochlea are then constructed by graph search techniques that gives the shortest path connecting all the landmarks. Because the paths are computed independently and in close proximity, sometimes they overlap or cross. As a post-processing step, manual edits to some of the paths are required. Example results of this process are in Fig. 1a.

3 Methods

We start the methods section with an overview of the proposed approach, followed by subsections providing more detail regarding novel components of the work. There are approximately 30,000 ANFs in a healthy human cochlea [12]. We represent them using auditory nerve bundles that are segmented along the length of the cochlea as shown in Fig. 1a [11]. To reduce the computational cost of our approach, we represent only 75 distinct bundles, each represents potentially hundreds of fibers. Our proposed nerve bundle action potential model is $P_M HM + P_U H(1 - M)$, where P_M and P_U are the action potential responses of single ANF cell biological nerve models (see Sect. 2.1) for a myelinated fiber and the degenerated, unmyelinated fiber model, respectively. H is the number of living fibers in the bundle that can be recruited for stimulation. M is the fraction, among those ANFs, of healthy versus degenerated ones. Thus, the bundle action potential is the superposition of the two fiber model action potential predictions scaled by the number of such fibers we estimate to be present in the bundle. We have designed an approach, described below, to determine patient customized values for these two parameters for each of the 75 distinct bundles.

The biological ANF model permits simulating action potentials (APs) created by ANFs as a result of the EF the ANF is subjected to. The EF sampled at discrete locations along the fiber bundle – each node of Ranvier (black nodes between myelinated segments in Fig. 1b) – is used to drive the ANF activation model. The EF generated by the CI electrodes can drive the ANF models and can be estimated using our CT-based high-resolution EAM of the electrically stimulated cochlea as described in Sect. 2.2.

Next, we will use our bundle model to simulate neural response measurements that can be clinically acquired. These measurements include recordings acquired using the CI electrodes of the combined AP signal that is created by the set of ANFs activated following a stimulation pulse created by the CI. Such measurements are called electrically evoked compound action potentials (eCAPs). Several eCAP-based functions can be clinically acquired. The most common are the amplitude growth function (AGF), which samples how the magnitude of recorded eCAPs (μV) grow as the current is increased for the stimulation pulse signal; and the spread of excitation (SOE) function, which measures the fraction of eCAP responses for two stimulating electrodes that are generated from the same ANFs [9, 10]. Both AGFs and SOEs can be simulated using our models and clinically measured using the patient's implant. While both AGF and SOE are rich with information about the electro-neural interface and have been acquirable for CI patients for decades, these metrics are not routinely used for clinical programming because they have been difficult to interpret. Thus, the method we propose provides a unique opportunity to (1) estimate neural health by tuning model neural health parameters so that model predicted eCAP functions match clinically measured ones; and (2) provide a physical explanation for the AGF and SOE measurements. Both of these typically unknown quantities could significantly improve an audiologist's ability to program the CI.

We tune neural health parameters for each ANF bundle so that simulated AGF functions for each electrode in the array best match the corresponding clinically measured ones. Finally, we conduct a validation study in which we evaluate our health prediction by simulating SOE functions using the model with the estimated neural health parameters and compare the results to clinical measured SOE to demonstrate the predictive value of our proposed models. The following subsections detail each step of our approach.

3.1 Dataset

N = 8 patients who had undergone CI surgery were used to create neural health estimation models. All the patients underwent pre- and post-implantation CT imaging needed to localize the intra-cochlear position of the electrodes and to create the tissue classification maps for the EAM models. The three clinical electrophysiological measurements critical for tuning and evaluating our models (EFI, AGF, and SOE) were also collected for all electrodes, for all patients with institutional review board approval.

3.2 Nerve Model

For each nerve fiber model, we follow the approach of Rattay *et al.* as we described in Sect. 2.1. We also used the same electrical and geometrical properties as Rattay did in his work [11]. The modeling is done using the NEURON simulation environment [15]. The overview of the auditory nerve fiber used in this study is shown in Fig. 1b. As shown in the figure, each nerve model consists of three subunits which are the peripheral axon, the soma and the central axon. The peripheral axon is located near hair cells in a human cochlea. They are myelinated when the fiber is healthy and fully functional. It is also common in patients with hearing loss that fibers where the peripheral axon has become unmyelinated exist and could have a weaker response to stimulation [14]. We define them as functional but 'unhealthy' ANFs. Then we can parameterize the health of each nerve bundle by varying the number of fibers, H, as well as the ratio of myelinated vs unmyelinated fibers, M, for each ANF bundle.

Our bundle model simulates bundle APs to the estimated EF generated by CI electrodes as previously discussed. Subsequently, eCAP measurements can be simulated in the model. To do this, each node of Ranvier for each bundle is treated as a current source, and the same finite difference method in Sect. 2.2 for estimating EF created by the CI is repurposed for estimating the EF created by the APs generated by all the bundles. This is done by defining bundle nodes as current sources corresponding to cross-membrane current. Thus, the result of each bundle model drives a new EAM to estimate the EF created by the ANFs in the cochlea. The value of the EF is then recorded at the site where the recording electrode is located. This process directly simulates the clinical eCAP measurement process.

In summary, the eCAP simulation can be divided into three steps: (1) for a given stimulating electrode, we calculate the EF using an EAM and record the resulting EF at the nodes of Ranvier for each nerve bundle; (2) we use those voltages as input to the neural activation models for both myelinated and unmyelinated nerves to compute our combined nerve bundle AP; and (3) we estimate the EF created by the bundle APs using another EAM, permitting simulated eCAP measurement from the position of recording electrode. In practice, in the final step an EAM can be created independently for each bundle and the compound response at the recording electrode is then given by

$$\text{simulated eCAP} = \sum_{i=1}^{75} P_{M,i}H_iM_i + P_{U,i}H_i(1 - M_i) \tag{8}$$

where $P_{M,i}$ and $P_{U,i}$ represent the value of the EF sampled at the recording electrode for the simulated eCAP of the myelinated and unmyelinated ANF model in the ith nerve bundle, respectively, and H_i and M_i are the number of fibers and fraction of those fibers that are healthy for the ith nerve bundle.

3.3 Optimization Process

Spoendlin *et al.* [12] found that for a healthy human cochlea, the average number of fibers can vary between 500 fibers per millimeter (mm) to 1400 fiber per mm depending on the location within the cochlea. Given that a nerve bundle in our model can represent a region as wide as 0.4 mm along the length of cochlea, we have set the boundary values for number of functional nerve fibers to be between 0 (all unresponsive) and 550 (all responsive) and the healthy ratio or the myelination ratio from 0 (all responsive nerve fibers are damaged) to 1 (all responsive nerve fibers are healthy).

Algorithm 1. Estimate the patient specific neural health parameters

Input: $\mathbf{P_{AGF}}$ = Patient AGF measurement
Variables: $\mathbf{S_{AGF}}$ = Simulated AGF data, **H** = Number of nerve fibers within bundles, **M** = Myelination ratio of fibers within bundles
Output: **HC** = Fiber count assigned to each control point, **MC** = Myelination ratio assigned to each control point
Start: Assign threshold and maxIteration, randomly assign **HC** and **MC**
While $\Delta|error|$> threshold and counter < maxIteration
  Interpolate **H** and **M** using **HC** and **MC**
 Calculate $\mathbf{S_{AGF}}$ using **H** and **M**
 For each electrode i
  $error_{AGF}$ [i] = mean(abs($\mathbf{P_{AGF}}$ [i] - $\mathbf{S_{AGF}}$ [i]))
 $error$ = mean($error_{AGF}$)
  Optimize **HC** and **MC** using a constrained nonlinear search based on Nelder-Mead simplex

Instead of determining values for H_i and M_i for each of the 75 nerve bundles independently, a set of control points are used to enforce spatial consistency in parameter values. We define n + 1 control points along the length of cochlea, where n is the total number of active electrodes. The control points are positioned to bracket each electrode. The parameters at those control points were randomly initialized with H_i between 0 to 550 and M_i from 0 to 1. The parameters for each nerve bundle are then linearly interpolated along the length of the cochlea using the control points.

We use the bounded Nelder-Mead simplex optimization algorithm [13] to optimize values at the control points. The cost function is calculated as the mean absolute difference between the simulated and measured AGF values for each electrode. Starting from a random initialization at each control point, our algorithm will iteratively calculate the parameters of every nerve bundle by interpolating control point values, simulate AGF using those parameters to evaluate the cost function discussed above, and determine new control point parameters using the Nelder-Mead simplex method until a maximum iteration number is reached or the change in error falls below the termination threshold (0.1 μV). Algorithm pseudocode is presented in Algorithm 1.

In our implementation, AGF values that were less than 35 μV were not included in the optimization process because low AGF values tend to be below the noise floor and are usually excluded from clinical analyses. During our experiments, Algorithm 1 is executed from 250 different random initializations for each patient model. The final fiber count and healthy ratio for every nerve bundle are determined as the median values across the 10 optimization runs that resulted in the lowest average error. This procedure diminishes the likelihood of choosing sub-optimal parameters that are local minima.

4 Results

The average absolute differences between the simulated and measured AGF and SOE values for fully customized EAMs are shown on the left side of Table 1. The average absolute difference between the simulated and the measured AGF values could be interpreted as the training error. Mann-Whitney U tests reveal significant improvement in AGF errors after training ($p < 0.01$). The error between the simulated and the measured SOE can be interpreted as the testing error since SOE was not used to optimize neural health parameters. Further, SOE is likely more sensitive to neural health than AGF because it is much more dependent on the spatial distribution of ANFs that contribute to the neural responses. The average SOE error across all patients after optimizing neural health parameters using our proposed method is 39.5 μV.

Table 1. Average mean absolute difference between simulated and measured AGF and SOE.

Subject #	Fully customized models			Generic models	
	AGF error – before optimiz. health (μV)	AGF error – after optimiz. health (μV)	SOE error-testing error (μV)	AGF error – after optimiz. health (μV)	SOE error-testing error (μV)
1	58	16	31	22	53
2	187	19	32	48	49
3	299	39	37	28	76
4	66	37	44	39	102
5	131	11	29	19	56
6	97	8	21	15	36
7	62	17	48	–	–
8	141	26	59	–	–
Average	134	21.6	39.5	28.5	62.0

In Fig. 3, we plot the simulation and clinical result of both AGF and SOE for subject 1. Both of the quantitative and qualitative comparisons show excellent agreement between neural stimulation responses that are clinically measured and those that are predicted by our parameter optimized models. We further compare the difference between neural health estimation using our fully customized models vs. generic models, where default electrical properties are used, for the first 5 subjects in the right side of Table 1. The AGF error (training error) resulting from the generic and electrically customized models is similar while the testing error with fully customized models is much smaller than generic models. A one sided Mann-Whitney U test reveals significantly better ($p < 0.05$) testing error with the fully customized model compared to the generic models. Example plots demonstrating the superiority of SOE simulations using customized for one subject are shown in Fig. 4. These results imply our patient-specific EAMs are critical, not only for EFI simulation, but also for accurate neural health estimation. An example neural health estimation result is shown in Fig. 1a, where the neural health color-codes are a combined function of both health parameters equal to $H\,(0.5 + M)$. Varying health of several regions of nerves was identified by the proposed method in order for prediction to match measured AGF.

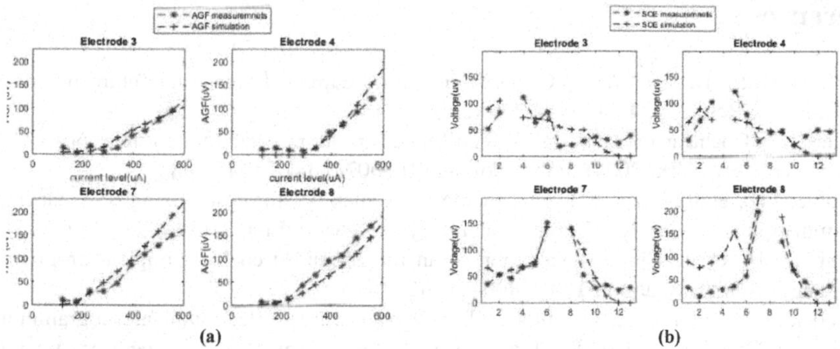

Fig. 3. (a) Comparison between measured and simulated AGF data. (**b**) Comparison between measured and simulated SOE data.

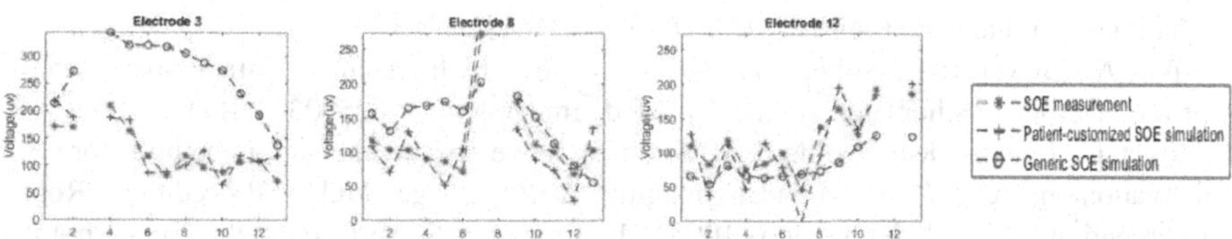

Fig. 4. SOE testing error for patient-customized versus generic models for Subject 4.

5 Conclusion

In this research, we developed an approach to estimate the health of ANFs using patient-customized, image-based computational models of CI stimulation. The resulting health parameters provide an estimate of the health of ANF bundles. It is impossible to directly measure the number of healthy ANFs *in vivo* to validate our estimates, however experiments with 8 subjects show promising model prediction accuracy, with excellent agreement between neural stimulation responses that are clinically measured and those that are predicted by our parameter optimized models. These results suggest our modelling approach may provide accurate estimation of ANF health for CI users. With the current IGCIP approach, assumptions are made about electrical current spread to estimate which fiber groups are activated based on their distance to the electrode. Our estimation on the health of ANFs may improve our estimation of neural stimulation patterns and lead to highly customized IGCIP strategies for patients. Our future work includes evaluating effectiveness of novel patient-customized programming strategies that use these models. Further, our methods could provide an unprecedented window into the health of the inner ear, opening the door for studying population variability and intra-subject neural health dynamics.

Acknowledgements. This work was supported in part by grant R01DC014037 from the National Institute for Deafness and Other Communication Disorders. The content is solely the responsibility of the authors and does not necessarily represent the official views of this institute.

References

1. Wilson, B.S., Dorman, M.F.: Cochlear implants: current designs and future possibilities. J. Rehabil. Res. Dev. **45**(5), 695–730 (2008)
2. Clark, G.: Cochlear implants. In: Speech Processing in the Auditory System, pp. 422–462. Springer, New York (2004). https://doi.org/10.1007/0-387-21575-1_8
3. Noble, J.H., et al.: Image-guidance enables new methods for customizing cochlear implant stimulation strategies. IEEE Trans. Neural Syst. Rehabil. Eng. **21**(5), 820–829 (2013)
4. Noble, J.H., et al.: Clinical evaluation of an image-guided cochlear implant programming strategy. Audiol. Neurotol. **19**(6), 400–411 (2014)
5. Noble, J.H., et al.: Initial results with image-guided cochlear implant programming in children. Otol. Neurotol.: Official Publ. Am. Otol. Soc. Am. Neurotol. Soc. Eur. Acad. Otol. Neurotol. **37**(2), e63 (2016)
6. Cakir, A., Dawant, B.M., Noble, J.H.: Development of a CT-based patient-specific model of the electrically stimulated cochlea. In: Descoteaux, M., Maier-Hein, L., Franz, A., Jannin, P., Collins, D.L., Duchesne, S. (eds.) MICCAI 2017. LNCS, vol. 10433, pp. 773–780. Springer, Cham (2017). https://doi.org/10.1007/978-3-319-66182-7_88
7. Cakir, A., Dwyer, R.T., Noble, J.H.: Evaluation of a high-resolution patient-specific model of the electrically stimulated cochlea. J. Med. Imaging **4**(2), 025003 (2017)
8. Ahmet, C., Labadie, R.F., Noble, J.H.: Auditory nerve fiber segmentation methods for neural activation modeling. In: Medical Imaging 2019: Image-Guided Procedures, Robotic Interventions, and Modeling, vol. 10951. International Society for Optics and Photonics (2019)
9. Briaire, J.J., Frijns, J.H.M.: Unraveling the electrically evoked compound action potential. Hearing Res. **205**(1–2), 143–156 (2005)
10. Hughes, M.: Objective Measures in Cochlear Implants. Plural Publishing (2012)
11. Rattay, F., Lutter, P., Felix, H.: A model of the electrically excited human cochlear neuron: I. Contribution of neural substructures to the generation and propagation of spikes. Hear. Res. **153**(1-2), 43–63 (2001)
12. Spoendlin, H., Schrott, A.: Analysis of the human auditory nerve. Hear. Res. **43**(1), 25–38 (1989)
13. D'Errico, J.: fminsearchbnd, fminsearchcon (https://www.mathworks.com/matlabcentral/fileexchange/8277-fminsearchbnd-fminsearchcon), MATLAB Central File Exchange. Accessed 16 Aug 2018
14. Nadol Jr., J.B., Young, Y.-S., Glynn, R.J.: Survival of spiral ganglion cells in profound sensorineural hearing loss: implications for cochlear implantation. Ann. Otol. Rhinol. Laryngol. **98**(6), 411–416 (1989)
15. Carnevale, N.T., Michael, L.H.: The NEURON Book. Cambridge University Press, Cambridge (2006)
16. Cartee, L.A.: Evaluation of a model of the cochlear neural membrane. II: comparison of model and physiological measures of membrane properties measured in response to intrameatal electrical stimulation. Hear. Res. **146**(1–2), 153–166 (2000). (14)
17. Malherbe, T.K., Hanekom, T., Hanekom, J.J.: Constructing a three-dimensional electrical model of a living cochlear implant user's cochlea. Int. J. Numer. Method Biomed. Eng. **32** (7) (2016). https://doi.org/10.1002/cnm.2751

Author Index

Ağıldere, A. Muhteşem 80
Akin, Oguz 80
Al Khalil, Yasmina 68
Amirrajab, Sina 68
Aparici, Fernando 41
Arbeláez, Pablo 120
Atkinson, David 110

Bomzon, Zeev 60
Borges, Pedro 110
Borotikar, Bhushan 90
Bourgeat, Pierrick 11
Bradley, Andrew 11
Breeuwer, Marcel 68
Burdin, Valérie 90
Butman, John A. 1

Cakir, Ahmet 184
Carass, Aaron 21
Cardoso, M. Jorge 110
Castillo, Angela 120
Chen, Baoying 50
Chen, Yiwei 142
Chou, Yi-Yu 1
Cook, Gary 110
Coşkun, Mehmet 80
Coupé, Pierrick 41

de la Iglesia-Vaya, Maria 41
Deshpande, Srijay 164
Dewey, Blake E. 1, 21
Douglas, Tania S. 90
Duzgol, Cihan 80

Escobar, Maria 120

Feng, Jun 50
Fookes, Clinton 11
Fouefack, Jean-Rassaire 90
Fripp, Jurgen 11

Gadermayr, Michael 131
Goh, Vicky 110

Haberal, K. Murat 80
Hawkins-Daarud, Andrea 32
He, Yi 142
He, Yufan 21
Hutton, Brian 110

Jackson, Pamela R. 32

Kläser, Kerstin 110
Kong, Wen 142

Li, Wanyue 142
Lin, Zi 101
Liu, Jingya 80
Liu, Ziteng 184
Lorenz, Cristian 68

Manjón, José V. 41
Minhas, Fayyaz 164
Modat, Marc 110
Mutsvangwa, Tinashe E. M. 90

Nbonsou Tegang, Nicolas H. 90
Noble, Jack H. 184

Ourselin, Sébastien 110

Pan, Xiaoying 50
Pham, Dzung L. 1
Prince, Jerry L. 21

Rajpoot, Nasir 164
Ranzini, Marta 110
Reich, Daniel S. 1
Reinhold, Jacob C. 21
Romero, Andrés 120
Romero, José E. 41
Roy, Snehashis 1
Rubio, Gregorio 41
Rusak, Filip 11

Salvado, Olivier 11
Santa Cruz, Rodrigo 11

Shamir, Reuben R. 60
Shao, Muhan 21
Shaw, Richard 110
Shi, GuoHua 142
Skočaj, Danijel 174
Stegmaier, Johannes 153
Štepec, Dejan 174
Swanson, Kristin R. 32

Thielemans, Kris 110
Tian, Yingli 80
Tourdias, Thomas 41
Traub, Manuel 153

Uhl, Andreas 131

Vécsei, Andreas 131
Vivo-Hernando, Roberto 41

Wang, Hongyu 50
Wang, Jing 142
Weese, Jürgen 68
Wimmer, Georg 131

Yang, Di 50
Ye, Chuyang 101

Zeng, Xiangzhu 101
Zhong, Manli 101
Zuo, Lianrui 21

Printed in the United States
By Bookmasters